TEACHER STRIKE!

TEACHER STRIKE!

Public Education and the Making
of a New American Political Order

JON SHELTON

**UNIVERSITY OF
ILLINOIS PRESS**
Urbana, Chicago, and Springfield

"'Against the Public': The Pittsburgh Teachers Strike of 1975–1976 and the Crisis of the Labor-Liberal Coalition" was originally published in *LABOR: Studies in Working-Class History of the Americas,* vol. 10:2: 55–75. Copyright 2013, Duke University Press. Republished by permission of the publisher.

A portion of chapter 2 was originally published under the title "Letters to the Essex County Penitentiary: David Selden and the Fracturing of America." *Journal of Social History* (Fall 2014) 48 (1): 135–155. Republished by permission of the publisher.

Library of Congress Cataloging-in-Publication Data
Names: Shelton, Jon, 1978– author.
Title: Teacher strike! : public education and the making of a new American political
    order / Jon Shelton.
Description: Second edition. | Urbana : University of Illinois Press, 2017. | Series:
    Working class in American history | Includes bibliographical references and index.
Identifiers: LCCN 2016043301 (print) | LCCN 2017000896 (ebook) | ISBN
    9780252040870 (hardback) | ISBN 9780252082368 (paper) | ISBN 9780252099373
    (e-book) | ISBN 9780252099373 (E-book)
Subjects: LCSH: Strikes and lockouts—Teachers—United States—History—20th
    century. | Teachers' unions—United States—History—20th century. | Public
    schools—United States—History—20th century. | Collective bargaining—
    Teachers—United States. | Labor movement—United States—History—20th
    century. | United States—Politics and government—20th century. | BISAC:
    EDUCATION / History. | POLITICAL SCIENCE / Labor & Industrial Relations. |
    HISTORY / United States / 20th Century.
Classification: LCC LB2844.47.U6 S54 2017 (print) | LCC LB2844.47.U6 (ebook) | DDC
    331.892/813711—dc23
LC record available at https://lccn.loc.gov/2016043301

For Kristina

CONTENTS

ACKNOWLEDGMENTS

All authors who write a book say they couldn't have done it alone. But I didn't realize why they take such pains to point out this fact until I finally finished my own. I have been incredibly fortunate to have received so much support, guidance, and friendship during the course of the very long time it took to produce this book. One of the great pleasures of my life has been getting to know the many people who helped me along the way.

I would have never even considered becoming an academic or writing a book if not for the students of Mariana Bracetti Academy in Philadelphia. I taught that school's first graduating class, and, as a teacher in the early No Child Left Behind era, became interested in how the structure of American political economy and education policy created such high hurdles for my students to overcome. Although I like to think I helped to educate my students, the education was much more formative for me. I could list all of my students, but I would like to thank, in particular, Sade Dancy, Rosemary Duarte, Fernando Flaquer, Jamaal Green, Kamia Hoilett, the late Sideem Ingram, John McGlone, Isaiah Muneer, Mario Pagan, Tim Pagan, Monique Ramos, Wayne Thornton, and Richard Torres.

I also would like to thank all of the archivists and librarians who make history possible. I am grateful to the staff at the Urban Archives at Temple University, the Pennsylvania State Archives in Harrisburg, the Heinz Historical Center Archives in Pittsburgh, the Robert F. Wagner Labor Archives at New York University, the George Meany Archives at the University of Maryland, and the LaGuardia and Wagner Archives at LaGuardia Community College. I did so much of the research and writing for this book at the Library of Congress in Washington, DC, that I probably should have moved in there at some point. I'd like to thank all of the reference librarians and staff who put up with my (many) requests. I have to offer an extra-special thanks to David Baugh at the City of Philadelphia Archives for helping navigate some challenging issues, and the folks at the Walter Reuther Library at Wayne State University, especially Dan Golodner, who helped me immensely and even showed me stuff that wasn't processed yet.

Financial support was indispensable: A fellowship from the National Academy of Education/Spencer Foundation helped fund an early version of this project and, more important, hooked me up with several great mentors. I also received a Grant-in-Aid of Research from the University of Wisconsin–Green Bay (UWGB), so thanks to the UWGB Research Council. I also appreciate the Labor and Working-Class History Association, especially the work of the committee (Alice Kessler-Harris, Michael Pierce, and Heather Thompson) who awarded me the Herbert Gutman Dissertation Prize. A summer stipend from the National Endowment for the Humanities proved crucial as I furiously overhauled the manuscript while everyone else I knew enjoyed vacations.

I have had too many smart people to count give me feedback on parts of this project over the years. I especially want to thank the Newberry Labor History Seminar in Chicago and the Havens Center for the Study of Social Justice at the University of Wisconsin–Madison for allowing me to workshop my research. Although this list may not be exhaustive, I appreciate the probing questions and insightful suggestions of the following people: Tom Alter, Patrick Barrett, Eileen Boris, Stephen Brier, Tess Bundy, Jennifer Burns, Jefferson Cowie, Diana D'Amico, Rosemary Feurer, Erik Gellman, Andrew Highsmith, Joseph Hower, Adam Mertz, Bill Reese, Kate Rousmaniere, Judith Stein, Elizabeth Todd-Breland, Micah Uetricht, Naomi Williams, and Lane Windham. Thanks also to the anonymous readers for the *Journal of Social History* and *Labor: Studies in Working-Class History of the Americas* and for the University of Illinois Press. Alice Kessler-Harris generously provided me with timely and helpful feedback at a difficult stage of the process. I am eternally grateful to Joe McCartin for his extensive and instrumental advice at a formative stage of this project, and to Will Jones, who has patiently read the entire manuscript more than once and asked me to answer difficult questions that have made this book exponentially better. Jerry Podair has been unbelievably generous with his time in addition to giving me crucial intellectual and professional support regarding this project. More important, our discussions about major league baseball helped keep me sane.

Thank you to the *Journal of Social History* and *Labor: Studies in Working-Class History of the Americas* for allowing me to reprint previously published material. Many thanks to Laurie Matheson, my acquisitions editor at the University of Illinois Press, who shepherded me through the production process, and to James Engelhardt, who took over the Working Class in American History series at an important moment and provided both enthusiasm and helpful advice. I am also very appreciative for the excellent and patient copyediting work done by Anne Rogers.

I owe a debt of gratitude to many people from the University of Maryland, College Park, who provided important mentoring: Ira Berlin, James Gilbert, Clare Lyons, Jennifer King Rice, and David Sartorius. David Freund and Robyn Muncy gave me important suggestions at a germinal stage of this project. I learned so much about how to think about history from Saverio Giovacchini. And without Julie Greene's sui generis combination of tough critique and warm support, this book would not be here. She is both a great mentor and terrific person.

I've been lucky to have a number of fantastic friends over the years, both inside and outside academia. Some helped me talk through my ideas and read drafts, but mostly they were just cool people who reminded me that other stuff happened in the world that didn't involve politics in the 1970s. Thanks to Colin Church, Eli Helman, Nate Matlock, Aaron McKenzie, Adam Parrillo, Dan Richter, Tim Sharma, and Ira Wigley. Special thanks to Stephen Duncan, who has been an insightful critic and friend.

At UWGB, I have had a number of students in the classes I teach ask fantastic questions and write papers that helped me clarify my own arguments. I especially want to thank the students of my spring 2015 U.S. Labor and the Working Class: Past and Present seminar: Paul Ahrens, Ben Freeman, Marcus Grignon, Eric Mandel, Tyler Rubenzer, and Sierra Spaulding. I am also fortunate to have the best colleagues in the whole world (and I'm not exaggerating). For support and intellectual stimulation, I thank Greg Aldrete, Caroline Boswell, Marcelo Cruz, Christin DePouw, Scott Furlong, Clif Ganyard, Doreen Higgins, Derek Jeffreys, JP Leary, Vince Lowery, Gabriel Saxton-Ruiz, Heidi Sherman, and David Voelker. Katie Stilp and Chris Terrien have helped me in indispensable ways. My unit—Democracy and Justice Studies—has given me intellectual support and, most important, friendship. I can't offer enough thanks to my friends Kris Coulter, Yunsun Huh, Katia Levintova, Eric Morgan, Kim Reilly, and Alison Staudinger. Thanks to Andrew Austin, the hardest-workin' program chair there is, and a truly generous person. Special thanks to the inimitable Harvey Kaye, whose mentorship and friendship know no bounds.

Finally, my family: Justin Sandherr, Evan Sandherr, and Tom and Tina Sandherr have supported me so much since I've become part of their family. Thanks to each of you! Steve Shelton was the first person I ever knew who was an intellectual, and I hope he knows how much he means to me. I'm convinced that my grandparents Lester and Erline are the kindest people who have ever lived, and I appreciate their support over so many years. Thanks very much also to my sisters, Angi and Arryn, and my brother, Stephen. I owe my father—Keith Shelton—more than I can say with words.

Finally, to my partner and our kids: I thought having two children would make it more difficult to complete a book, but the contrary was true. Trying to find as much time as possible to spend with two little people who make me so proud enabled me to work a lot harder when I needed to. Thank you, Sara and Keith. To Kristina, thank you for making every day of my life with you just a little bit better than the last one. You've leaped every hurdle with me along the way, and this book is just as much yours as it is mine.

TEACHER STRIKE!

From Labor Liberalism
to Neoliberalism

In September 1981, Philadelphia's 22,000 teachers went on strike and shut down most of the schools in the city. The mayor, Democrat William Green III, had demanded the layoffs of 3,500 teachers and that those left on the payroll give up a negotiated raise to pay for the schools' massive budget deficit. Teachers refused, and, after a two-month stalemate, the strike ended only when a Pennsylvania court ordered the teachers back to work under the conditions of the last contract; teachers lost their raises for good, but the layoffs were averted. This wasn't the first time the city's teachers shut down the schools for a lengthy period of time in a labor dispute. In the 1972–73 school year, for example, teachers walked out—in two separate strikes—for almost three months. In 1980, teachers in Philly were on strike again for three weeks. Philadelphia wasn't the only city where teachers went on strike, either. The 1981 strike, in fact, represented the bookend of a decade and a half of intense conflict between unionized teachers and the school districts of many of the largest American cities. Beginning with a one-day strike led by the United Federation of Teachers (UFT) in New York in November 1960, there were 300 teacher strikes in the 1960s—105 in 1967 alone. Strike action was even more pronounced in the 1970s, peaking at more than 200 in the 1975–76 school year.[1] Further, most of the longest and most prominent strikes occurred in the nation's major cities—New York City, Chicago, Pittsburgh, Detroit—and some, like those in Philly, lasted for weeks or even months at a time.

In addition, these cities represented the core of the labor-liberal coalition that structured public policy in the United States since the New Deal, and labor conflict in these urban education systems brought into dialogue many different constituencies of this coalition—teachers, civil rights activists, blue-collar workers, liberal elites, and corporate leaders. During crises like that of the Philadelphia school system in the 1970s, these Americans argued with each other, and many rethought their political assumptions. As a consequence, space opened for new ones. Indeed, conflict over urban education in the form of teacher strikes in the late 1960s and 1970s represents an important part of the explanation for why many Americans lost faith in the twin pillars of labor-liberal policy: the efficacy of government intervention in economic and social life and the necessity of a robust labor movement.

Teacher Strike! argues that this conflict over urban education in the United States was instrumental in cleaving the labor-liberal coalition along an axis in which one side—consisting of many white working- and middle-class Americans and corporate interests—imagined themselves as the only legitimate, "productive" contributors to society. This "producerist" coalition increasingly argued that both public employees and the urban poor flouted the law and siphoned off the resources of hardworking Americans, in their eyes causing the negative trajectory of political and economic events ailing the United States in the late 1960s, '70s, and early '80s. In fact, the critical attempts to understand what had caused the teacher strikes in the cities that had been bulwarks of the labor-liberal coalition helped to mainstream political assumptions that envisioned much more limited state action and the diminished role of labor unions. These assumptions undergirded the neoliberalism beginning to ascend by the late 1970s.

Clearly, the shift in American politics during the late 1960s, 1970s, and early 1980s—or what historians sometimes call the "long 1970s"—is a complicated story, and historians have worked for years to understand it. At this point, it seems indisputable that the long 1970s represented the era in which the New Deal order irrevocably lost traction. The growing salience of libertarian economics; mounting importance of the antiunion political economy and evangelical Christianity of the Sunbelt; reaction against feminism, the civil rights movement, and the counterculture; and a defense of the idea of America's greatness following Watergate and defeat in Vietnam combined to push the Republican Party to the right. In conjunction with the seeming failure of Keynesian macroeconomics to solve simultaneous economic downturn, energy crisis, and rampant inflation, the Democratic Party's move toward lifestyle issues and fecklessness in helping American workers in the wake of widespread deindustrialization did a good deal of the work necessary to shift the nation's political center from the New Deal labor liberalism that had promised to share American prosperity widely for generations.[2]

But a fuller explanation for this change necessitates a broad examination of the relationship between conflict over urban education and the nation's overarching political trajectory. Fundamentally, the struggles over urban education represented the confluence of the two major political interventions believed to be necessary from the 1930s to the 1960s for bringing economic democracy and prosperity to the United States: strong labor unions and robust government programs. A philosophical commitment on the part of the framers of the National Labor Relations Act (NLRA) to increase the purchasing power of workers combined with worker demands for greater industrial democracy to push the federal government to guarantee workers' rights to organize in 1935. Other policies of the New Deal (and its Great Society extension) expanded the role of the state dramatically. After the 1930s, the federal government guaranteed a minimum wage and subsidized mortgages, and federal/state/local partnerships distributed old-age insurance, Aid to Dependent Children (ADC—changed in 1962 to Aid to Families with Dependent Children, or AFDC), and, later, Medicare and Medicaid benefits. States and localities spent much more money on public education after World War II, and the federal government contributed compensatory funding for education beginning in 1965. Indeed, total public elementary and secondary school spending increased from $5.8 billion in 1949–50 to more than $70 billion by 1975–76 (a factor of nearly six when adjusted for inflation).[3]

At all levels, then, a significantly larger role for the government in promoting social welfare was evident, and a critical assessment of postwar labor liberalism should not neglect to point to the accomplishments that made the lives of so many Americans better. The labor-liberal coalition provided greater access to economic opportunity, more workplace democracy through unions, and a social safety net when the job market failed to provide a minimum level of economic security. Before the passage of revolutionary private-sector labor law that institutionalized the rights of workers to form labor unions, workers and employers literally fought things out in the street. After, labor violence virtually ceased, and although the benefits accrued disproportionately to whites and to men, many Americans worked in more dignified conditions, expanded the middle class, and enjoyed an unprecedented level of stability.[4] Thanks to increased state and federal funding and social welfare policies like the GI bill, education for Americans at all levels in the years after World War II reached unprecedented proportions. By the end of the 1960s, after decades of civil rights activism, the federal government had finally gotten around to enforcing the rights of African Americans and other minorities to equal protection of the laws and to participate in the political process. Indeed, as historian Harvey Kaye has argued, this generation of Americans was clearly the most progressive in U.S. history.[5]

Struggles over urban education in the long 1970s, however, prominently exposed three unique but interlocking series of limitations to the promise of New Deal labor liberalism. First, multiple levels of government failed in ensuring equal opportunity for all Americans—as civil rights activists in cities like New York, Chicago, and Philadelphia pointed out, many African American students in the postwar era attended poorly funded segregated schools, black teachers and principals faced institutional impediments to employment, and many parents in the inner city believed they sent their children to schools in which the teachers were unresponsive to their concerns. Inequality in education was a key component of a larger, systematic regime of segregated housing and limited economic opportunities in the nation's cities.

Second, the massive expansion of state services dramatically increased the number of public employees, and these employees' demands led to substantial labor conflict in the public sector. By 1960, there were almost 2 million teachers in the United States, for example, and that number reached almost 2.5 million by the mid-1970s.[6] The growing number of teachers helped expand access to education after World War II, but teachers also wanted the same rights enjoyed by their contemporaries in the private sector. They organized, and they went on strike for higher salaries and better working conditions, but when they did, school districts failed to ensure the balance of workers' rights and labor peace that New Deal private-sector law often achieved.

Finally, school systems could not remain financially solvent after fiscal crisis wracked American cities in the 1970s. The idiosyncratic system of American federalism treated states, counties, and cities as autonomous entities even though people and industries often crossed boundaries seeking jobs, investments, and lower taxes. In some cases, liberal policies like the Federal Housing Administration (FHA) and GI bill mortgage backing encouraged movement outside of cities; when homeowners and jobs alike moved elsewhere, city administrations were still largely responsible for the people who remained. In the tough economic climate of the 1970s, many cities struggled to maintain social services; their employees, especially teachers, were stuck trying to hold onto gains they had made after the advent of collective bargaining.

These three problems came together to lead to lengthy and intractable teacher-labor conflict in the 1960s, '70s, and '80s. Further, they caused many Americans to question the very underpinnings of a political system that proved incapable of finding viable solutions to these problems. The answers that emerged were different for many of the different constituencies of the labor-liberal coalition, and these discrepancies helped rend it asunder. Indeed, the cities in which urban education conflict took place in the most spectacular fashion became hothouses in which stalwarts of the New Deal coalition—from African American civil rights activists to white blue-collar workers—engaged in contentious public discussions that shifted the terrain on which Americans understood their political allegiances. Many white middle- and working-class Americans turned against labor liberalism, and tensions driving apart those who remained in the fold made it more difficult to defend it. These new divisions in American politics, in turn, opened space for powerful corporate interests, abetted by free-market ideology, to radically alter the trajectory of American political economy.

The Limitations of the Labor-Liberal Alliance

During the New Deal, the Roosevelt administration and its allies in Congress attempted to solve the problems of the Great Depression through legislation—first, through the National Industrial Recovery Act (NIRA), which the Supreme Court declared unconstitutional, and then the NLRA—that not only gave workers collective-bargaining rights but encouraged them to organize. Indeed, the opening section of the NLRA (or Wagner Act) argued that workers' lack of agency had helped cause economic downturn: "The inequality of bargaining power between employees who do not possess full freedom of association or actual liberty of contract and employers who are organized in the corporate or other forms of ownership association substantially burdens and affects the flow of commerce, and tends to aggravate recurrent business depressions, by depressing wage rates and the purchasing power of wage earners in industry and by preventing the stabilization of competitive wage rates and working conditions within and between industries." Such language may not represent the rhetorical flourish of "Workers of the World Unite!," but the policy inspired organizing drives in the 1930s among industrial workers and represented a far-reaching departure from the federal government's previous modus operandi toward unions: either specifically siding against workers (as in the 1894 Pullman strike) or lukewarmly supporting union efforts during periods of emergency only to prevent radicalism (as the Wilson administration did during World War I).[7] Further, the NLRA's architect—Senator Robert Wagner of New York—believed that collective bargaining would not only reignite economic growth but also represented "a system of checks and balances based on countervailing power." Beyond its necessity for economic democracy, then, Wagner also argued that "the development of a partnership between industry and labor in the solution of national problems is the indispensable complement of political democracy."[8]

Still, one should avoid romanticizing the regime of labor relations in the years following 1935. To begin, the 1947 revision of NLRA—the Taft-Hartley Act, passed over President Truman's veto by a Republican Congress—narrowed the ability of the labor movement to challenge both racism and corporate prerogative. The reactionary law withheld National Labor Relations Board (NLRB) protection from unions with elected leaders who refused to sign affidavits disavowing their connection to the Communist Party, restricted the tactics NLRB-certified unions could use against employers, and solidified individual states' abilities to weaken labor by allowing them to pass so-called right-to-work laws that prevented union security clauses in collectively bargained contracts. While organized labor's criticism of this "slave labor law" may have been somewhat histrionic, Taft-Hartley limited labor's ability to push beyond the shop floor for greater social democracy and seriously circumscribed its influence in the Sunbelt.[9] Further, Taft-Hartley's exclusion of supervisors from union rights allowed employers to prevent all manner of workers—from university professors to head nurses in hospitals—from collective bargaining.[10] Employers also developed methods to circumvent workers' rights still protected by the Wagner Act by the 1970s, and the law, by then, badly needed revision.[11]

As limited as the new legal protections may have been, however, deep changes in the American political culture occurred during the era of the labor-liberal coalition. There may have been no major party specifically devoted to labor as there was in Australia or the United Kingdom, but for the first time in American history, the large majority of wage laborers were in the fold of one party with a policy apparatus to support their economic aspirations.[12] As Steve Fraser and Gary Gerstle's now-classic formulation of the era puts it, the political order stemming from the New Deal "possessed an ideological character, a moral perspective, and a set of political relationships among policy elites, interest groups, and electoral constituencies that decidedly shaped American political life for forty years."[13] Indeed, combined with the Keynesian notion that the state should play a role in ensuring aggregate demand, the dominant political ideology for almost half a century centered on the belief that for the economy to function properly, wealth—to some degree—needed to be redistributed, and the state should play a role in fostering economic opportunity for all. Further, labor unions were integral in both the process of economic redistribution and in ensuring that workers enjoyed a political voice. Taken together, I call this belief, hegemonic in American politics from the New Deal until the late 1960s, *labor liberalism*.[14]

The Wagner Act, however, only applied to private-sector workers, and although these workers could organize, collectively bargain, and even strike, public-sector workers presented a problem. On the one hand, both the federal government and most state governments extended some version of bargaining rights to those whom they employed. In 1959, Wisconsin became the first state in the United States to broaden these rights to public employees, and in 1962, President John F. Kennedy issued Executive Order 10988, offering explicit support to workers employed by the federal government to enjoy union representation and collective bargaining. Following the president's lead and responding to organizing by public-sector workers, nine additional state governments joined Wisconsin and passed their own laws authorizing collective representation by 1966. By 1980, most states had passed laws allowing some level of public-sector bargaining rights.

On the other hand, while federal labor statutes assumed that workers sometimes needed to withhold their labor as leverage for negotiations, neither Kennedy's executive order nor many state laws allowed public employees to go on strike. State governments, in part, recognized public employees' rights to bargain in response to their organized demands for greater workplace democracy, but an equally significant reason for public-sector policies was actually to prevent the interruption of government services by strike. Thus, new state laws used a carrot-and-stick approach to curtail strikes. New York's 1967 collective-bargaining law was typical. While sanctioning the unionization of public-sector employees and providing an apparatus for mediation, the law also provided stringent penalties for striking.

Perhaps the best illustration of this ideal is the public-employee law passed in Pennsylvania in 1970. Two years earlier, in response to a strike by Pittsburgh teachers, a Republican governor convoked a commission to revise the law. The commission's recommendation formed the basis for a new policy that offered public-sector unions the right to strike as a "safety valve" to prevent strikes by public employees.[15] The commission believed that recalcitrant local governments would be more willing to bargain if they knew public employees had recourse to strike legally, which would actually prevent strikes in the long run. But this law failed, in large part because it gave local authorities a loophole to make strikes illegal. In short, the problem in Pennsylvania and elsewhere stemmed from the fact that public-sector strikes were illegal but arresting or replacing thousands of workers was virtually impossible.

Some voices did advocate rectifying this problem. President Nixon's undersecretary of labor called for full union rights for federal workers in 1974, and organized labor, led by the American Federation of State, County, and Municipal Employees (AFSCME) pushed for a "Wagner Act for Public Employees" for much of the 1970s. The U.S. Congress flirted with such a law—Representative William Clay (D-MO) introduced a bill to grant public employees the same rights that other workers enjoyed under NLRA, including the right to strike—but this effort was abandoned following the economic downturn of the mid-1970s and especially after the Supreme Court, citing the Tenth Amendment, struck down the federal government's jurisdiction over the working conditions of state and local employees in *National League of Cities v. Usery* (1976).[16]

The contradiction, then, became that although public employees in most states could join unions and collectively bargain, laws also barred them from using the major weapon at their disposal for exacting leverage in those negotiations. Teachers—a highly organized group of middle-class professionals who, unlike police officers, firefighters, or sanitation workers, were not immediately essential for ensuring public safety—represented an especially intractable problem for the liberal state.

This problem took on a significant gender dimension. Indeed, teachers were almost always forced to strike illegally, and because the teaching force consisted disproportionately of women—who since the nineteenth century had been charged with caring for children and moral guidance as much as teaching reading, writing, and arithmetic—their abrogation of the law signified to many Americans a growing disrespect for authority.[17] Large urban teaching forces represented a particular practical problem because effectively punishing them for striking was next to impossible. Although most of the local and national leadership of teacher unions were men, the challenge represented by breaking the law forced local school boards—consisting largely of men and gendered male in public discourse—either to arrest scores of female teachers (which still was not likely to end a strike) or to capitulate to the demands of women workers. Either course was subversive and, for many Americans, called into question the trajectory of the nation.

Thus, in the late 1960s and '70s, the uncertain legal position of public-sector unions as it related to teachers in the nation's largest cities represented one important reason for the insoluble problem of acrimonious shutdowns of urban education. This conflict often reached epidemic proportions. In the fall of 1975, for example, teachers walked out in New York City; Chicago; Boston; one-third of the cities and towns in Rhode Island; thirty-one school districts in Pennsylvania (including Allentown—the state's fourth-largest city);

Berkeley, California; Billings, Montana; Lynn, Massachusetts; Wilmington, Delaware; and New Haven, Connecticut. Two million school students missed time in September, and teachers in Lynn, New Haven, and Wilmington served jail time.[18] Teachers in Chicago struck nine times between 1968 and 1987.[19] Indeed, it is not an exaggeration to say that strikes by teachers became a key public-policy problem that cried out for dramatic change, much as the "labor problem"—conflict between repressive employers and striking workers leading to the shutdown of services like the railroads and destruction of property—had been in the late nineteenth century.[20]

The second intractable question on which teacher strikes shone a light is that many New Deal policies institutionalized racial inequality and failed to rectify it. This problem by the 1970s was clearly evident in the northern and midwestern cities of what constituted the geographic core of the labor-liberal coalition. Seeking economic opportunity and pushed out of rural labor by mechanization, African Americans moved from the South to cities across the United States in the 1940s and '50s as the federal government—through the FHA, GI bill, and other programs—facilitated homeownership in whites-only suburbs. In the cities, administrative forms of racial segregation in schools by local authorities held the color line even in places where it was illegal. Efforts to rectify the massive disparity of economic and education opportunities between blacks and whites—exacerbated by discriminatory federal policy—led to a divisive series of confrontations between civil rights advocates and a political system either unwilling or unable to repair the damage done by years of oppression and inequality.[21]

The history of public-sector employee unionization and race in the United States is complicated. Historians are just beginning to explore this topic, and much work remains to be done. Several developments, however, are clear. On the one hand, African Americans often faced less discrimination in the public sector, particularly in federal employment. African American postal workers, for example, fused efforts for civil rights and union rights and were instrumental in dramatically improving conditions for all postal workers in the 1970 U.S. postal wildcat strike.[22] As historian Joseph Hower has argued, the 1968 campaign in Memphis of striking black sanitation workers—immortalized by Martin Luther King Jr.'s assassination on April 4 at the Lorraine Motel—represented a seminal moment in which AFSCME became firmly linked to civil rights and built on this momentum to organize blue-collar African American public employees in the subsequent years.[23]

On the other hand, African Americans represented a major constituency receiving public services in American cities, and unionized public employees—disproportionately white in many cases—became, for some activists, the physical embodiment of the unequal opportunities blacks faced in segregated neighborhoods whose residents suffered from poverty. As Frances Fox Piven framed public-sector militancy in an essay from November 1969, in large cities like New York, the "keenest struggle is with residents of the central city ghettos (who in any case now form a substantial segment of the 'general public' in most cities). Policemen, fire-fighters, teachers and public welfare workers increasingly complain about 'harassment' in the ghettos. For their part, growing numbers of the black poor view police, firemen, teachers, public-welfare workers and other city employees as their oppressors."[24]

Urban teachers represented a particularly crucial group in this struggle. City teachers spent years struggling to be treated as professionals, and this professionalism centered on due-process rights and academic freedom. In the context of the failure of liberal policy to provide equality of opportunity between whites and blacks, however, Black Power activists demanded control over community institutions, and this control prominently included schools. Kwame Ture and Charles Hamilton's seminal 1967 book, *Black Power: The Politics of Liberation*, for example, called for a radical change in the way inner-city schools were run, arguing that "control of the ghetto schools must be taken out of the hands of 'professionals,' most of whom have long since demonstrated their insensitivity to the needs and problems of the black child. These 'experts' bring with them middle-class biases, unsuitable techniques and materials; they are at best dysfunctional and at worst destructive."[25] Community groups in New York City, Newark, Philadelphia, and elsewhere would answer Ture and Hamilton's call, leading to conflict with majority-white teacher unions.

Sometimes, gender played an important role in this conflict as well; the call by Black Power activists for a greater role for schools and for teachers in nurturing black children in the context of an oppressive system often led to increased demands for care work. For many female teachers—particularly in the elementary schools—the collective-bargaining era had ushered in a view of professionalism that meant freedom from the arbitrary demands of mostly male administrators and never-ending nonteaching duties like supervising students on the playground, in the lunchroom, and during transportation to and from school. Demands that both sides believed were necessary led to bitter conflict.

Finally, racial inequality within metropolitan areas and the resulting concentration of poverty combined with the idiosyncratic nature of American federalism to create another key problem by the 1970s: fiscal crisis. As historians such as Thomas Sugrue, Robert Self, and Jefferson Cowie have shown, deindustrialization began well before the 1970s. Indeed, the history of American industry after World War II can best be thought of as a protracted movement of capital and jobs: first to suburbs like Levittown (outside Philadelphia), then to the business-friendly South, and, finally, across the border.[26] Cities across the United States hemorrhaged industry and manufacturing jobs during the second half of the twentieth century, a situation that federal macroeconomic and trade policy did little to alleviate.[27] Further, the decentralized nature of government—in spite of the interdependent nature of metropolitan political economy, cities and suburbs were expected to be autonomous—in the United States led state and local governments to compete for businesses by lowering taxes, and overwhelmingly white suburban residents could benefit from the amenities of a city while paying lower taxes elsewhere. The loss of industry and affluent homeowners eroded urban tax bases and often led to massive budget deficits exacerbated by national and international economic turmoil in the 1970s.

Indeed, as the economy declined, and energy prices, inflation, and unemployment all increased simultaneously, many cities faced fiscal crises at the same time that more money was necessary to mitigate the even greater hardships faced by urban populations. Budget crises, then, in cities like Philadelphia and New York City in the 1970s and '80s—which often forced teachers to strike—resulted from the disconnect between the promises of liberalism to provide economic and social equality; the metropolitan spatial dynamic abetted by federal, state, and local policies; and the failure of a patchwork system of policies to alleviate structures of poverty.

The Limits of Labor Liberalism Lead to Conflict

In the 1960s and '70s, the tension around public-sector labor rights, racial discrimination entwined in liberal policy, and fiscal crisis led to three distinct waves of urban teacher strikes, distinguished by the primary problems driving each. In each instance, teachers' actions—often against the law—brought on a further unraveling of the labor-liberal assumptions undergirding urban education (as well as other facets of American politics). The first wave, initiated in a 1960 strike by New York City teachers, lasted until about 1968. These strikes came at the crest of almost a century of efforts on the part of organized teachers to gain decent salaries, more control over teaching conditions, and the right to collectively bargain. They did not represent the first teacher strikes in the United States—if not a strike exactly, teachers in

Chicago demonstrated against Depression-era budget cuts in the 1930s, and teachers staged significant strikes in St. Paul and Buffalo, for example, immediately after World War II—but the strikes in the 1960s were notable in that they succeeded in pushing school boards to establish lasting collective-bargaining relationships.

Given the brevity of these strikes, criticism of teachers and their defiance of the law was limited. That changed in the late 1960s and early '70s when a new wave of teacher conflict revolved around tension between Black Power activists and teacher unions, and white working- and middle-class reaction against both took the form of the proliferating discourse of law and order. Although teacher strikes for basic representation still occurred (in St. Louis, for instance, teachers still struggled for a basic bargaining relationship during a notable two-month strike in 1973), from 1968 until the economy began to spiral into downturn in 1973, conflict revolved mostly around civil rights activists' attempts to gain control of school personnel and resistance from largely white ethnic teaching forces in cities like New York City and Newark. Indeed, this conflict drove a major wedge between teachers and much of the African American community. Even in cities like Philadelphia—where strikes did not pit teachers as dramatically against Black Power—teachers and civil rights activists failed to forge social-movement coalitions against an urban political economy that did disservice to both groups.

Although teachers' labor rights and racial conflict still factored into teacher strikes, from 1973 forward, dramatic, illegal teacher actions revolved around drastic fiscal crises. The decimated tax base, continued loss of jobs, and movement of wealth to the suburbs combined with economic downturn and the higher salaries of city teachers in the collective-bargaining era to create massive budget shortfalls. During the late 1960s, banking interests had been willing to cover deficits with credit, but in the 1970s they increasingly moved to discipline cities with austerity budgets that hurt both public employees and the urban citizens who both relied on those services and had been critical of the subpar versions that had existed before fiscal crisis. Although the new political economy of the city would hurt most public employees, the poorest communities were hardest hit, and minority teachers (some of the last hired) were disproportionately hurt by layoffs.

The Philadelphia teacher strike of 1972–73 served as the forerunner of this last wave of strikes. Resulting from a budget deficit that the board of education tried to alleviate by cutting positions and lengthening the school day, the yearlong conflict kept teachers out of classrooms for three months in total. Working-class whites from the urban fringes allied with law-and-order mayor Frank Rizzo to take a hard line against teachers and oppose tax hikes that would have eased the financial conditions of the schools. Moreover, the public discourse around the strikes initially pointed to the unique problems associated with deindustrializing American cities and an urban-suburban political economy built on local property taxes. After a local judge enjoined the strike and sent teachers to jail, however, the driving narrative about the conflict revolved around the city's perceived inability to control the excessive demands of its own labor force and asked whether it was legitimate to spend money on a school system of mostly black students. Much of this posturing focused on the masculinist stance of Rizzo—who publicly argued that teachers wanted to make him and his followers into "patsies"—against any display of "weakness" toward the union.

Labor conflict brought about by the New York City fiscal crisis in 1975 represented the fulcrum that changed how cities' public sectors operated. The battleground for the crisis encompassed virtually every aspect of city life in that year, but the most prominent conflict involved striking teachers and a dramatic showdown over whether teachers would invest their pension funds in city paper. The yearlong debate—among New York's citizens, in the New York City media, the national media, and in Washington—focused on who to blame for the massive budget shortfall. For much of that year, loud voices attempted to make the case that the city faced an extreme iteration of the difficulties with which many American cities struggled by the mid-1970s. By the time the federal government provided a loan guarantee in late 1975, however, the explanation for the city's crisis revolved around the intransigence of the city's "nonproducers": its unionized workforce and the welfare recipients who many argued were unwilling to work.

Some of this link likely hinged on the gender of teachers. Although more men had entered the profession by the early 1970s—the female percentage of the nation's teachers had declined to about 65 percent—and although UFT president Albert Shanker took much of the heat in concert with other male labor leaders in the city, it is hard to ignore the connection between a largely female workforce and a social-welfare program—AFDC—that was clearly understood by many white working- and middle-class Americans as a gendered, and a racialized, program.[28]

After 1975, it was clear that metropolitan or federal solutions were off the table, and other cities were chastened by the new reality. Even in cities like Pittsburgh and St. Louis—where subpar salaries and benefits for public employees in part had kept the cities fiscally solvent—the specter of becoming "another New York City" gave school boards the wherewithal to resist striking teachers for a long time, leading to protracted conflict there in 1975–76 and in 1979, respectively. By September 1981, during the last lengthy teacher strike in American history—in Philadelphia, just after President Reagan replaced striking air traffic controllers—fiscal crisis forced teachers into either striking illegally yet again or giving back things that had already been negotiated in a signed contract.

What I hope to show by telling this story is how many ordinary Americans viewed the labor-liberal coalition through the prism of teacher strikes. Each of three different series of conflicts, layered on top of each other, led many Americans representing the critical bulwark of the labor-liberal coalition to view the entire enterprise as less viable, and, for some, to place responsibility for the problem on the very people—like teachers and urban African Americans—caught in the crosshairs of institutional failure.

The Special Importance of Education

The significance of education in the long 1970s is not a surprising development. As a fundamental facet of everyday life as well as the single-most expensive expenditure of local governments, the manner in which the state provides education to its citizens has been a major political battleground for much of American history. Indeed, schools and their teachers have long been charged with an overwhelming sense of importance. Historian William Reese has best characterized this long-standing phenomenon, arguing that "citizens who should know better routinely expect them to accomplish what is humanly impossible, complain bitterly when the schools falter, and yet turn to them again and again to cure social ills not of their making." [29]

The significance charged to schools and teachers became even more pronounced after World War II, when federal, state, and local investment in education increased dramatically, partially in recognition of the fact that education had become even more fundamental for Americans in procuring opportunity as the U.S. economy transitioned toward more highly skilled, white-collar employment. In addition, education became, in the postwar years, perhaps the primary battleground over race, gender, poverty and inequality. To sum up these developments, it is worth quoting from the Supreme Court's unanimous decision overturning segregation in *Brown v. Board of Education* in 1954:

> Today, education is perhaps the most important function of state and local governments. Compulsory school attendance laws and the great expenditures for education both demonstrate our recognition of the importance of education to our democratic society. It is required in the performance of our most basic public responsibilities, even service in the armed forces. It is the very foundation

of good citizenship. Today it is a principal instrument in awakening the child to cultural values, in preparing him for later professional training, and in helping him to adjust normally to his environment. In these days, it is doubtful that any child may reasonably be expected to succeed in life if he is denied the opportunity of an education. Such an opportunity, where the state has undertaken to provide it, is a right which must be made available to all on equal terms.[30]

Indeed, conflict in the education sector was of fundamental interest because of the central place education played in so many social conflicts in the postwar world. By the time teachers went on strike in massive numbers in the late 1960s and '70s, the civil rights movement had focused on the public-education system for decades. *Brown v. Board of Education* and the massive resistance it provoked from southern whites hinged on who had access to education and on what terms. Clashes like those at Arkansas's Central High School in 1957 or the bloody battle in 1962 to integrate Ole Miss involved the federal government in literal fights to ensure that everyone could access education. Desegregation became a major battlefield in the North as well by the end of the 1960s, and the busing of school students to achieve integration was a hot-button issue in national politics over the course of the long 1970s. The 1960s campus protest movements—including the struggles of Chicanos and African Americans for ethnic studies programs and other institutional reforms—also revolved around the role education played in the trajectory of the nation. Critics of student radicals often wondered why their teachers had failed to inspire the appropriate respect for authority and hard work. When teachers went on strike, then, these associations almost reflexively became a part of the public conversation, thus heightening an already tense sense of crisis.

In addition to exposing racial animosities and spotlighting the spatial privileges enjoyed by many urban and suburban whites, teacher strikes also disrupted assumptions about gender in postwar America. The earliest efforts by teachers to unionize in the late nineteenth century consciously embraced feminism and powerfully connected salaries and pensions for female teachers to tax collection and municipal budgets in cities like Chicago.[31] Thus, teacher organization had long been controversial in part because it "feminized" control over public finances. Going back to battles over teacher unionization in the early twentieth century, critics of Chicago teacher activist Margaret Haley and city schools superintendent (and National Education Association [NEA] president) Ella Flagg Young chastised the Chicago school district for its "frenzied feminine finance."[32]

In the late 1960s and '70s, the revival of feminism, the increase in two-parent working families, and the challenges to the male breadwinner ideal helped to galvanize a substantially gendered opposition to a largely female workforce's fight for higher salaries and dignity in the workplace.[33] As Alice Kessler-Harris has argued, "ideals of service and sacrifice to children, parents, and spouse have often been thought to render liberty for women irrelevant and women's search for individual rights 'selfish.'"[34] During the era of the teacher strike, the push by teachers for collective rights was also routinely characterized as "selfish" because of the necessity of care work in the profession, and teacher power was seen as inherently subversive. Indeed, male school board representatives and other critics of teacher unions often used the language of emasculation at the expense of city leaders to describe teachers' efforts for higher salaries in the 1960s and '70s, and in the age of fiscal crisis, union activists wondered why it appeared that teachers were asked to make greater sacrifices than traditionally male occupations like the police or firefighters. Some male blue-collar workers also resented the higher status that accrued to white-collar teachers and were especially galled when economic conditions got tougher in the 1970s that their tax dollars seemed to support higher salaries for teachers.[35]

Teasing out just how much the gendered nature of the teaching profession affected calls for austerity, however, is complicated. It is the case that a majority of teachers were female in the decade, but for much of the postwar era, the profession included increasing numbers of men. At perhaps the crux of teacher militancy, in 1971, in fact, the percentage of men in the nation's teaching force was higher than at any other time in the twentieth century. Further, during some calls for austerity, predominately female teaching forces were linked to male-dominated public-employee unions like police, firefighters, and sanitation workers. In addition, it is difficult to say for sure whether the profession became more feminized in the late 1970s—a process that has continued largely unabated until the present—as a consequence, but it seems likely that calls for austerity strengthened the gendering of the profession, particularly given that salaries of teachers remained stagnant from the early 1970s until the 1980s.

In any event, teacher strikes meant so much because teachers were clear representatives of the state. Certainly, all public-sector workers represent the government in some sense, but perhaps no other worker personified the state more directly than a teacher because a teacher's labor—as a literal government intervention—would be visible in one's public school education and in parents' interactions with their children's teachers. Moreover, as Paul Johnston has shown, public-sector workers' labor disputes inherently revolved around public policy. In the case of teachers, this meant education policy—how much the state should spend on educating students as well as how and what to teach them.[36]

Furthermore, strikes by teachers worked to alter the public image of labor as much as strikes of any other groups of workers because they directly impacted the everyday lives of Americans in a way that strikes by other workers did not. The late 1960s and early '70s represented an era of rank-and-file militancy by private-sector workers in the United States, and so teacher strikes were not the only ones in the decade.[37] A strike by factory workers, however, more directly harmed their employers—by curtailing their profit-generating ability—than the public, who could continue to purchase goods from the competitor of a struck company. Even strikes by other public employees, like sanitation workers, did not immediately affect the public in a way that teacher strikes did, as it might take days, for instance, for garbage to pile up. Teacher strikes, indeed, clearly spotlighted an often-neglected feature of the education system: it is a massive child-care subsidy necessary for many parents of elementary-age students that is typically hidden from public consideration. When the women who cared for children in the public school system walked out, families were forced to reckon with the necessity of finding alternative forms of care.

Finally, teacher strikes were so significant because of the greater amount of dollars spent on education in the postwar era, the greater complication of the funding structure of schools, and the connections Americans made between these developments and their own tax burden. For much of American history, the federal government contributed virtually nothing to public education, and states provided very little. Although most state constitutions guarantee public education, the administration of schools for much of American history took place on the local level. In the Progressive Era, state governments became more involved in setting statewide goals and providing grants-in-aid to local districts. Still, in the 1929–30 school year, state governments provided less than 20 percent of school funding, and local governments funded the rest. These percentages shifted in the 1930s as cash-strapped local governments turned to states for help and shifted even more in the postwar era as states became more interested in solving issues of social equity through education spending. States provided almost 40 percent of school funding by the 1940s–'60s, and by the 1978–79 school year, for the first time, accounted for more education funding than local governments.[38] Further, following the USSR's launch of *Sputnik* in 1957, Congress passed the National Defense Education Act (1958), which made the federal government a player in education spending, a role that increased dramatically through compensatory spending for poor school districts with the passage of the Elementary and Secondary Education Act (ESEA) in 1965. Federal funds for public elementary and secondary schools leaped from $650 million in the 1959–60 school year to $3.2 billion in 1969–70 and $9.5 billion in 1979–80.[39]

Total education spending on all levels increased dramatically in the 1950s, '60s, and '70s. When adjusted for inflation, total spending increased 49 percent in the 1950s, 73 percent in the 1960s, and 20 percent in the 1970s.[40] Taking these trends—increased total spending and a greater diffusion of costs among American citizens through additional spending at the state and federal level—in tandem, it is evident why education policy became so highly politicized. For many Americans squeezed in the 1970s by the poor economy and a tax burden made more onerous by an inflation rate over 7 percent for the decade, education became an obvious political target.

The New Order

Following the decline of the labor-liberal coalition's hegemony in American politics during the course of the 1970s, it was far from inevitable that the political center in the United States would shift right. Other alternatives were possible, even by the end of the decade. After gaining huge majorities in Congress following the Republicans' Watergate debacle in 1974 and regaining the presidency in 1976, Democrats perhaps had an opportunity to revivify the labor-liberal coalition. The two most important moments in this possible trajectory were the Humphrey-Hawkins Act, which began as a bill to guarantee full employment but ended up being watered down to the point of uselessness when signed into law in 1978, and the Labor Reform Act of 1977—a bill to make it more difficult for employers to circumvent the NLRA—which died by filibuster in June 1978. One could imagine that guaranteeing jobs for all Americans and making it easier for workers to organize and collectively bargain might have seriously reshaped the economic and political trajectory of the United States in the 1980s and beyond. For historian Judith Stein, President Jimmy Carter looms large in the failure of these two policies, as he put little effort into what were basically congressional initiatives and instead pushed tax cuts for the wealthy and tight monetary policy (appointing chairman of the Federal Reserve Paul Volcker, who raised interest rates to astronomical levels in a tunnel-visioned effort to end inflation). Jefferson Cowie's account of the late 1970s argues that Carter lacked not only the will but also "the political winds to drive innovative policy" as labor liberalism had been irrevocably fractured earlier in the decade.[41]

Whatever the case, the decline of the labor-liberal coalition—a decline that included both the failures of the state at the national level and the metropolitan-level failures represented by the teacher strikes that I document—took place at the same time that an increasingly coordinated, high-caliber assault on labor unions and the social-welfare state from both corporations and free-market ideologues was gaining a foothold in the American political mainstream. As Gérard Duménil and Dominique Lévy have argued, finance capital in the United States and Europe responded to a critical diminution of profits in the 1970s by pushing to discipline labor and open up capital movement across the globe. To Duménil and Lévy, this "neoliberalism" expressed "the desire of a class of capitalist owners and the institutions in which their power is concentrated, which we collectively call 'finance,' to restore—in the context of a general decline in popular struggles—the class's revenues and power, which had diminished since the Great Depression and World War II. Far from being inevitable, this was a political action."[42]

Attacks on social democracy by financial interests, as in the case of the New York City fiscal crisis, were indeed fundamental in this project. But these efforts also represented part of a longer-term struggle to roll back and reshape New Deal policy from its very inception. A network of intellectuals who opposed the interventions of the New Deal when it was unpopular to do so formed the vanguard of this effort. Kim Phillips-Fein, for example, has shown how the Mont Pelerin Society, formed by libertarian philosopher Friedrich von Hayek in 1947 and supported by American business interests, brought together free-market economists and other intellectuals from the Austrian school and the University of Chicago to challenge labor-liberal hegemony.[43]

The most prominent neoliberal intellectual in the United States by the late 1960s—Chicago economist Milton Friedman—emerged from the Mont Pelerin milieu. Friedman was not a tool of class interests, and during the course of several decades developed a coherent critique of the interventions in the marketplace represented by the New Deal state and labor unions. Friedman viewed himself as a "liberal" in the mold of Adam Smith, who had argued in the influential text *The Wealth of Nations* (1776) against the arbitrary power of the sovereign in driving economic production for the benefit of the crown. Like Smith, the University of Chicago professor envisioned a diminished role for the state: "The scope of government must be limited. Its major function must be to protect our freedom both from the enemies outside our gates and our fellow-citizens: to preserve law and order, to enforce private contracts, to foster competitive markets." Also, Friedman argued that the modern "liberal" must also push for dramatic change: "The nineteenth-century liberal was a radical, both in the etymological sense of going to the root of the matter, and in the political sense of favoring major changes in social institutions. So too must be his modern heir."[44]

Indeed, Friedman advocated limits on all manner of government programs—he had argued, for example, that the New Deal had been entirely unnecessary because the Great Depression resulted from flawed monetary policy, and as an advisor to President Nixon, had been instrumental in ending military conscription. Friedman envisioned the nation's education sector functioning like a marketplace; while he didn't argue against using public funds for schools, he believed that parents should be given vouchers to shop for the school of their choice.[45] The economist's views gained even wider political currency in 1976 after he won the Nobel Prize for Economics, especially since his prize-winning work dealt with inflation, a particularly acute problem in the American economy by the mid-1970s.

Although Friedman envisioned privatization and market competition as neutral developments, geographer David Harvey reminds us that the "actual practices of neoliberalization" in the years since the 1970s represent something very different. Indeed, a global crisis in capital accumulation in the 1970s was solved, he argues, not by making the economy more productive through competition but through "accumulation by dispossession." Many of the institutions built for the public good instead became assets for private capital: "Privatization (of social housing, telecommunications, transportation, water, etc. in Britain for example), has in recent years, opened up vast fields for overaccumulated capital to seize upon. . . . If capitalism has been experiencing chronic difficulty of overaccumulation since 1973, then the neo-liberal project of privatization of everything makes a lot of sense as one way to solve the problem."[46] Similarly, Steve Fraser has argued that the "political economy of auto-cannibalism" has driven life in the United States for the past forty years: "Over the last generation, public facilities, resources, and services—waterworks, transportation, public lands, telecommunication networks, airwaves, herbs, forests, minerals, rivers, prisons, public housing, schools, health care institutions, in sum everything from zoos to war-fighting—got privatized, turning the commercial into profit centers for the incorporated."[47]

For schools, as Pauline Lipman's groundbreaking work on Chicago has shown, "actually existing neoliberalism" there led to major subsidies to corporate, real estate, and banking interests at the expense of the city's most impoverished students: "The neoliberal state is noninterventionist when it comes to regulating capital (except when stability of the system is at stake) and providing for social welfare, but interventionist when it comes to ensuring the rule of markets and favorable conditions for capital accumulation." Lipman documents how by the 1990s, Chicago politicians were using tax increment financing (TIF) to divest funds earmarked for schools in low-income neighborhoods for the benefit of private economic development elsewhere.[48]

Detailed studies of other metropolitan areas in the vein of Lipman's work on Chicago must be done to fully map the shape of neoliberalism in the years since the 1970s. And there is still much more to be done to fully understand the roots of neoliberalism elsewhere; as Elizabeth Tandy Shermer's recent work on postwar Phoenix has shown, a "developmental 'neoliberalism'" emerged in the Sunbelt even before it had begun to take hold in cities like New York and Chicago in the late 1970s.[49] Further, the privatization that has more firmly grasped the U.S. economy in the years since the decline of labor liberalism in the 1970s was first imposed in extreme fashion in Chile, following the violent overthrow of the democratically elected regime of Salvador Allende in 1973. More efforts to situate the emergence of neoliberalism outside the United States and to show how those stories interact with the U.S. context are needed to fully understand the phenomenon.[50]

From what we know now, however, it is apparent that while there may very well be a contradiction between theory and practice, neoliberalism as an ideology structuring political assumptions in the United States has gained much wider purchase since the 1970s. As I understand neoliberalism, David Harvey's definition best characterizes its underlying logic. To Harvey, neoliberalism represents "a theory of political economic practices that proposes that human well-being can best be advanced by liberating individual entrepreneurial freedoms and skills within an institutional framework characterized by strong private property rights, free markets, and free trade."[51] The role of the state in this formulation is not just to stay out of the market, as Adam Smith argued, but also to actively facilitate new market opportunities for private capital accumulation.

Neoliberal ideology involves a good deal of intellectual work to be persuasive. In part, it is imbued with legitimacy because, to use former British prime minister Margaret Thatcher's often-used words, "there is no alternative." As private-sector labor unions have struggled for survival and public-sector unions struggle to resist privatization and all that that entails (the charter school movement in the United States has emerged as a source of profits for private management companies in some states, for instance, and everywhere has created a largely nonunion alternative teaching force), resistance to neoliberalism is difficult and involves a good deal of personal risk. On the other hand, however, the developments of this book highlight the importance of ideology in the growing consent among many Americans for neoliberalism in the years since the collapse of the labor-liberal coalition.

Neoliberalism relies on the notion that virtually every aspect of life is better off organized by a marketplace because the "competition" sorts out the winners from the losers. Discourses of "productivity," then, are fundamental in the logic of neoliberalism. In fact, the growing political connection in the 1970s between corporate America and working- and middle-class whites around an imagined dividing line between the "productive" and the unionized teachers and urban poor who supposedly siphoned off their tax dollars appears to have been an important precondition for neoliberal hegemony.[52] Further, these moves hinged on long-held assumptions about race (as urban teachers—white or black—worked disproportionately with African American students) and gender (both teachers and AFDC recipients were often coded as female in public discourse) that only accelerated in the years since the events documented in *Teacher Strike!* As Harvey has argued, neoliberalism relies in part on appeals to "cultural nationalism" to demonize "'liberals' who had used excessive state power to provide for special groups (blacks, women, environmentalists, etc.)."[53] Similarly, Lipman argues that characterizing African Americans and Latino/as in Chicago as undeserving, lazy, and unproductive has been a crucial part of the neoliberal project of privatization.[54]

To put it simply, corporate interests that promoted arguments about production in the marketplace as a means for pushing privatization and disinvestment from public goods needed to be able to argue that existing public investments were used fruitlessly and wastefully. To capitalize on political formulations conjoining teachers and impoverished students who were unable to transcend the limitations of diminished economic opportunity in the deindustrializing city, then, was a logical move. Public discussions of teachers and the students they taught—particularly after cities faced fiscal crisis in the 1970s—shifted from a vision of education as an aspect of life necessary for democracy and potentially able to ameliorate poverty in the city to one obsessively focused on whether teachers were of any "value" to the productive "taxpayer." In fact, by the Philadelphia strike of 1981, critics in the public sphere had begun to promote vouchers as a means of disciplining teachers into productivity through the marketplace.

Outline of the Book

It is clear that the decline of the labor-liberal coalition and the emergence of neoliberalism in the years to follow is an exceedingly complex story. It is the story of how millions of Americans became convinced that the United States wasn't working as it should, and, eventually, either enthusiastically or by default, that increased competition—between "productive" and "nonproductive" people—was the answer. This book is my modest attempt to highlight a previously underexamined aspect of American life in the long 1970s to help understand how and why things changed so rapidly. Admittedly, this book is also ambitious in the sense that I bring together several different threads of analysis in a series of American cities during this time. By no means is it meant to be either a comprehensive history of any one city, nor is it a comprehensive account of all the many different teacher strikes in the United States.

It is also not a social history of teachers. Although I sometimes document the ways in which individual teachers and union leaders engaged in the public controversies brought about by strikes, more fine-grained analysis of the ways teachers—within their various race, class, and gender positions— experienced these conflicts is needed in the future. Among other questions, I believe such work will help us understand what it meant that a largely female workforce—typically fronted by male union leadership—was forced to bear the brunt of the blame for a set of political-economic conditions beyond their control. It is my hope that other scholars can build on the framework I've developed here.

Teacher Strike! uses local case studies to paint a larger picture of the connection between education, the public-sector labor movement, and political change in the 1960s and '70s. Important studies—like Jeffrey Mirel's now-classic book on Detroit—have examined individual school systems in the context of a single metropolitan region over a long period of time.[55] Other scholars have contributed valuable case studies of teacher unions during the postwar era; some of these, in particular Jerald Podair's *The Strike That Changed New York: Blacks, Whites, and the Ocean Hill–Brownsville Crisis*, have shown the connection between conflict over urban education and broader political change.[56] With the exception of Marjorie Murphy's groundbreaking work on the American Federation of Teachers (AFT) and the NEA, however, no other studies bring together the local and national politics around teacher unionization as this book does. My work builds on the longer history Murphy has charted to study the interactions between the political economy of urban schools, teacher unions, and national political developments in a way that focuses more closely on a particularly integral period of American history.[57]

Chapter 1 charts the early history of teacher unionization and efforts by states in the orbit of the labor-liberal coalition to deal with the growing militancy of teachers. In particular, it uses Pennsylvania's public-employee legislation in 1970 to understand the policy assumptions of labor liberalism toward public employees. By using the case studies of Philadelphia and Pittsburgh, I show how, by the early 1970s, the teacher strike became a key problem that the labor-liberal coalition proved incapable of solving.

Chapter 2 examines the conflict between advocates of Black Nationalism and teacher unions in the late 1960s and early 1970s. I show how strikes in Ocean Hill–Brownsville in 1968 and Newark in 1970 and 1971 deepened the rift between Black Power advocates and unionized teachers and solidified the teacher strike as a site of national political crisis. I also document how Black Power–teacher union conflict emerged in less spectacular fashion in several other major cities. The chapter concludes by using a series of letters written to AFT president David Selden in prison to understand how teacher labor conflict led many Americans to equate striking teachers with the racially charged category of the "urban rioter."

Chapter 3 focuses on a wave of teacher strikes in early 1973. In January and February, teachers simultaneously walked out in Philadelphia, Chicago, and St. Louis, prompting commentators to worry that the American education system was falling apart. I show how, in both Philly and St. Louis, an emerging narrative that these cities could not control their workforces made citizens in each metropolitan area question the legitimacy of labor liberalism. Focusing on debates about political economy, the chapter shows how white working-class Philadelphians in particular had begun to link both teachers and African American citizens as unproductive.

Chapter 4 examines the New York City fiscal crisis of 1975, with a focus on the role of the public-sector workforce, especially teachers. Unions—like the UFT in September and October of 1975—used aggressive tactics to mitigate some of the worst of the cuts proposed by New York's banking interests. In this effort they were somewhat successful, but these limited victories came at the expense of further empowering their opponents to argue that they invested in their own livelihoods at the expense of the city. The dominant explanation for the city's crisis increasingly revolved around the intransigence of the unionized workforce—particularly the teachers—and they were discursively linked with New York's welfare recipients as "unproductive" citizens who needed to be curtailed to save New York.

Chapter 5 focuses on a bitter two-month strike by the Pittsburgh Federation of Teachers in 1975–76. Just months after the darkest days of the crisis in New York, the strike led many in the Pittsburgh area to fear a similar set of circumstances if teachers were allowed pay raises that would keep pace with inflation. Through debate in the city's two major newspapers, a narrative emerged that pitted the interests of teachers squarely against the interests of the city's "productive" citizens. The fight galvanized a new Governor's Commission to revise Pennsylvania's public-employee law. This time, teacher unions were on the defensive, signifying how seriously postwar labor liberalism had been undermined.

Chapter 6 returns to St. Louis and Philadelphia for strikes in 1979 and in 1980–81, respectively. While St. Louis teachers won a modest contract victory following a grueling three-month strike, the public discussion cleaved the city's "productive" taxpayers from teachers and the disproportionately poor school-age population. The Philadelphia city schools were in desperate financial shape, and Mayor William Green, betting against public support for teachers, reneged on a contract with the city's teachers in order to get concessions. The second Philadelphia strike provides a marked contrast to the gains made by public-sector unions in the 1960s and '70s. Following the popularization of school vouchers as an alternative to the public school system in the late 1970s, calls for market-based solutions to the city's fiscal situation gives us a window into the emergence of the neoliberal alternative in the years to come.

In the conclusion, I chart the connection between teacher unions and American politics from the 1980s to the present and ask what kinds of political allegiances are possible in the neoliberal world that has followed.

CHAPTER I

"A New Era of Labor Relations"
Teachers and the Public-Sector Labor Problem

In March 1969, women's magazine *Redbook* addressed the growing phenomenon of teacher strikes. Surveying nationally prominent strikes from
1968, the story pointed to Florida, where, in the spring, 25,000 National
Education Association (NEA)–affiliated teachers resigned en masse with
the expectation of forcing the state to spend more money on public education, and to the series of United Federation of Teachers (UFT) strikes across
New York City over community control of school personnel in Brooklyn.
The article further raised the specter of a "possible *nationwide* walkout."
Taking the perspective of mothers, the piece asked, "Are teachers now more
concerned about money and power than they are about children?"[1] The
questions emerging from the article highlight the surging importance of
teacher unions' efforts to win collective-bargaining rights, salary increases,
and smaller class sizes in the series of crises gripping the nation in the late
1960s.

The growing conflict over teacher strikes in the late 1960s and early '70s
underscored new questions about the place of organized labor during this
time period. Though not without their critics, unions had represented a key
component of the post–World War II economic and social order.[2] At the same
time, most liberals—Americans holding that robust government intervention
was necessary to ameliorate social conflict and ensure widespread economic
prosperity—disapproved of strikes, particularly in the public sector. Indeed,
labor stoppages disrupted the flow of goods and services, as well as the
sense of peaceful, rational operation that characterized the postwar ideal
for government.[3]

This chapter charts the assumptions within American political culture
regarding public-sector unions in the late 1960s and early 1970s through an
examination of teachers. I show how organizing teachers, from their earliest

efforts to form unions, were steeped in controversy and struggled for much of the twentieth century to exercise collective power. At times, teacher unions pushed for radical versions of economic, racial, and gender equality, but many of these more wide-ranging possibilities were lost by the 1950s. In the 1960s, mainstream politicians—both Democrat and Republican—in many states comprising the foundation of the labor-liberal coalition had, by the late 1960s, extended some of the same rights to public-sector unions that the Wagner Act gave private-sector industrial workers in the 1930s. Liberal politicians did so in part because of the increased electoral usefulness of a growing public-sector workforce that could be effectively mobilized through unions. But they extended these rights mainly because public employees organized effectively and made demands. The earliest efforts of these work-ers included bold demonstrations of teacher power through short strikes. Further, teachers and other public employees successfully argued that they deserved the same rights as workers in the private sector. Indeed, in 1959, only 38 percent of Americans believed that public school teachers should be able to unionize, but that number jumped to 61 percent less than ten years later.[4]

Public employees, however, almost always lacked the right to strike. Unionized teachers in many places fought for bargaining power, and, like employees since the onset of the wage-labor relationship, encountered em-ployers not interested in ceding prerogatives over pay and working condi-tions. Thus, teachers had to strike for bargaining leverage, but in doing so, broke the law. In turn, many liberal state governments grappled with this emergent problem, trying to develop policies that could prevent illegal labor stoppages.

At first, the public discussion around these new policies highlighted a fundamental optimism about the efficacy of the state. Indeed, legislators, newspaper writers, and even outspoken citizens all believed that public-sector bargaining was vitally important in the rational administration of government services, and that the right policy could solve the problem of labor conflict. I use the passage of Pennsylvania's Public Employee Act 195 in 1970 to demonstrate the reach of these assumptions. In tracing the public discussion of the most radical policy effort to forestall public-sector strikes, I show that the liberal faith in rational policy was alive and well in the late 1960s.

But the new policies did not actually stop strikes, and this chapter ar-gues that because public-sector militancy—in particular, on the part of schoolteachers—emerged in so many places, a sense of crisis around the problem developed in the 1960s and early '70s. Again, I use Pennsylvania as a prominent example of this phenomenon by focusing on conflict in Philadelphia and Pittsburgh. As this chapter makes clear, teachers' low salaries and challenging working conditions elicited a great deal of sympathy for teacher unionization, but many Americans also became anxious about teachers—as an overwhelmingly female workforce—exerting power in the workplace by withholding their labor.

The Rise of Teacher Organization

Many studies of the labor movement in the United States during the 1970s emphasize the waning influence of the AFL-CIO (American Federation of Labor–Congress of Industrial Organizations) on federal policy, deindustrialization, the decline of militancy in the private sector, and the advent of concession bargaining.[5] A focus on the trajectory of public-sector unions, however, seriously complicates this declension narrative.[6] Examining the distinction between public- and private-sector unionization rates, for example, challenges the notion that private-sector labor power began to lose steam abruptly in the 1970s. In fact, the high-water mark of private-sector union density was in 1953. In that year, 15.5 million working Americans in the private sector belonged to unions, a number representing around 36 percent of those who worked for private employers. By 1962, however, that percentage had declined to around 32 percent, by 1973 to less than 27 percent, and by 1983 to just under 18 percent. The aggregate number of Americans in private-sector unions also declined to just over 13 million by 1983. Public-sector growth in the 1950s, '60s, and '70s helped to obscure this trend because it kept total unionization rates robust. From 1953 to 1983, the number of public-sector workers in unions increased sevenfold, from 770,000 to 5.4 million, and union density in the public sector increased from 12 percent to its peak, just above 40 percent, in 1974, tapering off to about 34 percent by the third year of the Reagan presidency.[7]

During the era of postwar public-sector organizing, teacher unionization represented a highly visible and controversial effort. This contentiousness is not surprising in light of the historical development of the profession in the United States. The public education system represented the first widespread social-welfare program in the United States; broadly available, publicly financed education began in the first half of the nineteenth century, and its reach was only possible because early architects of public education like social reformer Horace Mann could mobilize women to enter the teaching force at lower salaries than men. During the course of the 1800s, the teaching profession was gendered female, and an ancillary benefit—for those who believed common schools served the purposes of social control—of feminizing education meant that women's natural affinity for moral guidance ensured that new generations of children learned American republican values like respect for the nation, selflessness, obedience, and a strong work ethic.[8]

These expectations for teachers persisted well into the twentieth century, and much of the early history of teacher organization involved unionized teachers calling for radical economic and gender equality. Although professional organization in education dates to the mid-nineteenth century (the NEA was formed in 1857), the first serious attempts to organize teachers into unions came with the Chicago Teachers' Federation (CTF) in the 1890s. The CTF, led by president Catharine Goggin and firebrand vice president Margaret Haley, sought higher salaries and better working conditions through collective action, developed an explicitly feminist agenda, and forged connections with both the families of their students and the wider Chicago labor movement. Most impressively, Haley, in the early 1900s, connected teachers' salaries and pensions to Chicago's political economy by using an extensive study of the budget to show that some of the most prosperous corporations dramatically underpaid their taxes while the city claimed it was unable to increase teacher pay.[9]

The center of teacher organization, however, abruptly diverged from its feminist origins in the 1910s. In 1916, three Chicago locals joined teachers from Gary, Indiana, to form the American Federation of Teachers (AFT). Male high school teachers, however, took over the process and installed a male teacher—Charles Stillman of the Chicago Federation of Men Teachers, which was formed after seeing the success of the elementary school–focused CTF and was designed to push the concerns of the city's mostly male high school teachers—as president. In the words of historian Marjorie Murphy, the turn would "privileg[e] male leadership in the union at its inception. Although women presidents were to follow, the tone of the leadership would be set early: it was an affirmation of male abilities and in effect a rejection of the style and methods of the feminist leadership of the CTF." Under pressure from the Chicago school board, Haley and the CTF left the AFT; the CTF moved even further from the AFT after it officially supported World War I (Haley strongly opposed it).[10]

Though less consciously challenging gender norms, teacher unionization continued to be controversial. The Red Scare and attacks on industrial unions by corporations after World War I left organized labor reeling in the 1920s, making teacher unionization of any kind difficult. The 1919 Boston police strike, in which Massachusetts governor Calvin Coolidge famously rebuked the strikers that "there is no right to strike against the public safety by anybody, anywhere, any time" and replaced them, drew national attention and catapulted Coolidge into national politics. The police drew so much animosity that, afterward, AFL president Samuel Gompers repudiated all public-sector strikes.[11] Even after Congress passed the NLRA, public-sector strikes remained contentious. President Roosevelt, in 1937, did believe that, as it was for private-sector workers, organization in the public sector was "both natural and logical." He also asserted, however, that a "strike of public employees manifests nothing less than an intent on their part to prevent or obstruct the operations of government until their demands are satisfied. Such action, looking toward the paralysis of the government by those who have concern to support it, is unthinkable and intolerable."[12]

In this atmosphere, teacher organizations limited their goals. The NEA—which included both teachers and administrators and eschewed collective bargaining—focused on lobbying for national legislation supporting public education. NEA leaders viewed teachers embracing militant unionism as "unprofessional," and the AFT became what Murphy calls a "gadfly union" in the 1920s. The Depression years, however, she argues, transformed the AFT into the "bread-and-butter union that emerged in the early sixties."[13] Indeed, the distressed government budgets during the Great Depression galvanized many teachers, especially in New York City and Chicago. In the Windy City, teachers were paid in scrip or not at all and attacked banks downtown to demand paychecks. But as the labor movement grew during the 1930s, the AFT also moved "into the bosom" of the conservative AFL, leading to a confrontation about the direction of the national teacher union. In New York, many teachers were attracted to radical critiques of American capitalism during the Depression, and a struggle ensued between Communists and anti-Communist liberals over control of the Teachers Union—AFT Local 5. Although the AFT initially refused to expel the local, growing anti-Communist repression in the United States, coupled with Local 5's support for the Soviet Union following the Nazi-Soviet pact, led the national union to revoke its charter in 1941.[14] A new anti-Communist local—the Teachers Guild—took its place.[15]

Immediately after World War II, teacher walkouts inflamed controversy over public-sector unions. In 1946–47, American industrial workers undertook a massive strike wave. Teachers across the United States, who had not seen the pay increases of their counterparts in the private sector during the war, went on strike, too. Teachers won higher salaries in an NEA affiliate-led walkout in Norwalk, Connecticut, and teachers in AFT locals went on strike for higher salaries in St. Paul, Minnesota, in 1946 and in San Francisco, and

Buffalo in 1947.[16] Just as private-sector workers faced a backlash in the form of Taft-Hartley, conservative legislatures passed state laws prescribing stiff penalties for strikes by government employees. The 1947 Buffalo strike—where teachers' salaries were lower than they had been in 1932 even though the cost of living had more than doubled—inspired the legislature to pass the most draconian of these.[17] New York's Condon-Wadlin Act (1947) was designed to unequivocally punish public-sector strikes. Under the law, teachers did not enjoy collective-bargaining rights, and if they went on strike, they could lose their job; if rehired, they stood to lose tenure rights and would be barred from raises for six years. Union leaders could also be stuck with fines and prison sentences.[18]

Just as Cold War paranoia—in the form of the Taft-Hartley requirements denying legitimacy to radical-led unions—limited the scope of private-sector labor activism, so the anti-Communist backlash in the postwar years helped to choke off more radical forms of social-movement teacher unionism in the late 1940s and '50s. Indeed, local, state, and federal authorities tried to connect unionized teachers to Communism. In New York City, in the late 1940s and 1950s, the school board suspended or fired teachers with "subversive" ties and ultimately banned the Teachers Union (TU).[19] Like other private-sector unions with radical ties, the CIO expelled the TU in 1950, and both the House Un-American Activities Committee (HUAC) and the Senate Internal Security Subcommittee (SISS) investigated the influence of Communist teachers in the schools.[20] Many of these teachers—especially Communist TU teachers—had been some of the staunchest advocates for racial equality and robust academic freedom.[21]

In no small part because of the limited alternatives, mainstream teacher unions by the 1950s had become much more narrowly oriented, focusing mostly on being treated as "professionals" and gaining better salaries and more control over working conditions. As Jonna Perrillo has shown, by the late 1950s, many white teachers resisted efforts by the New York City Board of Education to get them to teach in the city's highest-poverty, overwhelmingly African American and Puerto Rican schools. In fact, some teachers saw a few years in inferior schools as little more than a stepping stone to a better work climate (represented by less-challenging classrooms) and viewed efforts by the school board to equalize teacher quality as an assault on their professionalism.[22]

Because of the emphasis on professionalism in teaching—an ideal going back to nineteenth-century attempts to standardize and centralize education in urban school districts across the United States—many teachers viewed the notion of forming trade unions with ambivalence. On the one hand, many teachers believed they differed from the blue-collar workers who comprised most of the unions in the United States. On the other hand, the fact that unions had made private-sector workers much more prosperous began to eradicate many teachers' doubts. In 1955, for instance, the average New York City schoolteacher earned $66 a week—$6.35 less than one could make washing cars.[23] The professional ideal could also lead teachers to believe they required unions to ensure that school administrators treated them with respect. Nonunionized teachers in the 1950s and '60s, indeed, often felt humiliated when principals unilaterally assigned teachers to lunch and playground supervision or ignored workplace grievances. Albert Shanker, for example, recalled that at his first teaching job in an East Harlem high school, his principal chewed him out in front of his students for not properly cleaning up the classroom and assigned the future UFT and AFT president to monitor students at a local store during his lunch break to ensure they weren't stealing.[24]

Furthermore, teachers in the postwar era—hoping to gain the middle-class status implied by professionalism—placed a premium on job security. As Dan Lortie's sociological study of teachers in the 1960s and '70s has shown, the profession attracted "many persons who have undergone the uncertainties and deprivations of lower- and working-class life."[25] Thus, for those who became teachers, arguably the most important benefit of teaching was the occupation's stability, and unionization helped to ensure it. Indeed, before collectively bargained contracts, many school districts still forced female teachers—whether married or not—to inform their principal upon getting pregnant and expected them to resign, and, even where tenure existed, the school district largely controlled all personnel assignments.[26] Although many of these teachers were men seeking to secure a family wage, it is clear that the demands of female teachers for job security and livable salaries represented a challenge to dominant notions of gender and work.[27]

Many teachers—particularly in the nation's cities—supported racial equality. Indeed, the AFT staunchly supported the civil rights movement in the 1950s and '60s. Most dramatically, the AFT expelled its Atlanta local in 1955 when it refused to desegregate; sent activists to teach in Prince Edward County, Virginia, after local authorities closed down the schools rather than integrate; and prominently supported the March on Washington in 1963.[28] Further, teacher unionization in the 1960s corresponded with a new generation of teachers, some of whom were more conscious of inner-city poverty and wanted to improve both the schools and their working conditions. In Chicago, for example, historian John Lyons argues that "an often underestimated cause of the teacher militancy was generational discontent. Many of the most vociferous complaints about working conditions came from younger teachers who had entered the Chicago school system in increasing numbers as enrollments soared." Indeed, a new generation of teachers in the 1960s "exuded a rebellious spirit and declining respect for authority, and they rejected the conformity and complacency of the older generation."[29]

Still, these teachers represented a minority—if at times a vocal one—and teacher organization by the 1960s focused primarily on "bread-and-butter" concerns: the major goal by then, as it was for private-sector unions, was exclusive collective-bargaining rights in order to gain higher salaries and smaller class sizes as urban schools struggled to accommodate children whose families had moved to northern cities like New York and Chicago during and after World War II.

Teachers gained collective-bargaining rights, but only after organized struggle. Led by organizers like David Selden and Albert Shanker, the UFT—the merger of the anti-Communist Teachers Guild and other smaller teacher groups in 1960—staged a successful one-day strike just before the presidential election pitting John F. Kennedy against Richard Nixon. The city's mayor—Robert Wagner Jr., son of the architect of the NLRA—had effectively begun collective bargaining with some of the city's public employees after passing Executive Order 49—the so-called Little Wagner Act—in 1958. Therefore, although the UFT strike was illegal, it was neither feasible nor desirable to replace thousands of teachers, and the union won a representation election for the city's teachers the next year. A one-day strike by the UFT in 1962 forced the board of education to agree to the first collectively bargained, guaranteed contract for teachers in the United States.

This effort inspired teachers elsewhere and initiated an era in which school districts in many places ensured better salaries and working conditions for teachers.[30] Buoyed by organizing funds from the United Auto Workers (UAW)–backed Industrial Union Department of the AFL-CIO, AFT organizers helped to forge collective-bargaining arrangements in many other major U.S. cities; teachers in Detroit, Philadelphia, and Cleveland signed contracts by 1965, and in Chicago in 1966. Marjorie Murphy best sums up the meteoric nature of teacher organization during this era: by the end of the 1970s, "72 percent of all public school teachers were members of some form of union that represented them at the bargaining table. Before 1961, unions in less than a dozen school districts could claim they represented only a small fraction of schoolteachers."[31]

Still, teacher militancy (in the context of aggressive public-sector unionization) was not uncontroversial, even within the labor movement. AFL-CIO president George Meany's intercession with Mayor Wagner during the UFT

strike in 1960 had proven to be crucial; nonetheless, according to one in-
sider, the AFL-CIO "responded with caution" in the 1960s, as "members of
the traditional unions viewed the demands of the government workers from
their newly acquired perch in the middle class" and feared higher taxes. Public employees, however, made up the major source of growth for the labor
movement, and so the AFL-CIO's tentative support for public-sector unionization continued; Meany publicly supported Executive Order 10988 giving
bargaining rights to federal employees, and the AFL-CIO Executive Council
helped local unions push for state and local collective-bargaining statutes.

Much of organized labor, however, was wary about encouraging strikes;
just as liberal politicians feared public backlash, so did the AFL-CIO. Thus,
while the national organization nominally supported striking public-sector
workers—as the AFL-CIO did in paying the $220,000 worth of fines levied
on the UFT for the Ocean Hill–Brownsville strike and as the UAW and other
unions pushed for property-tax increases to pay for higher teacher salaries
in Detroit when teachers threatened to strike in 1965—behind the scenes
they sometimes discouraged strikes, fearing that a backlash "would limit
strikes in the private sector as well. [Leaders of craft and industrial unions]
did not want the legitimation of compulsory arbitration or labor courts,
both proposals for resolving public sector strikes. Once placed in operation, they might be a simple step to extend these proposals to the private
sector."[32] As we shall see, teacher strikes, indeed, emerged as a key problem
for the labor-liberal coalition in the late 1960s.

Teacher Strikes in the 1960s

New York City teachers were pioneers in the postwar era—necessarily using the illegal strike to gain exclusive representation rights, a collectively
bargained contract, and a dramatic raise by 1962. Other teachers followed
their example, and these efforts led to a handful of teacher strikes in 1964
and 1965. In November 1964, for example, 300 members of the Louisville
Federation of Teachers (LFT–AFT) went on strike for ten days to get higher
salaries. In May 1965, one-third of the teachers in South Bend, Indiana,
struck illegally for four days. New York City teachers threatened to strike
again that September, and the shutdown was averted the day before the new
school year began only when the city agreed to about $40 million in raises,
or approximately $800 for each teacher for each year of the two-year con-
tract. In December, 750 Newark teachers (members of the Newark Teachers
Union—AFT) walked out for two days, seeking an election for exclusive
representation. Union officers there were held in contempt of court.[33]

There were thirty documented teacher strikes in 1966. The AFT local in New Orleans went on strike, offering a challenge to a region (the South) that the NEA had long dominated. NEA locals became more militant in northern cities, too: the Newark Teachers Association (NTA), for instance, struck for two days in February. The growing militancy of the AFT had pushed the NEA to begin embracing collective bargaining, and, tentatively, supporting strikes by affiliates. (Before that, the organization's primary means of influencing working conditions had been to use "sanctions" to get local school boards and state legislatures to improve salaries and working conditions; NEA teachers were advised not to teach in sanctioned school districts). In the fall of 1966, teachers in Michigan—all AFT locals—defied the state's antistrike labor law, and teachers "boycotted" one hundred school districts in the Detroit suburbs and in Flint in a "militant bid for pay raises."[34]

In 1967, even more teachers walked out of classrooms for collectively bargained contracts and/or exclusive representation—there were 105 labor stoppages in all—and the strikes became longer and more contentious. Local authorities arrested 130 teachers in Baltimore that May, for example, as part of a two-day strike by 1,000 Baltimore Teachers Union (BTU–AFT) members. In September, according to one account, "turmoil gripped school systems in many areas of the country . . . as teachers struck or prepared to strike at the start of the school year." Indeed, teacher strikes impacted almost two million students that September: in East St. Louis, Illinois, 500 of the city's 800 teachers went on strike for higher salaries and recognition of their AFT local; a judge ordered them back to work. In addition, 15,000 teachers walked out in thirty-four Michigan cities, including 10,000 Detroit teachers seeking a $1,200 raise and a shorter school year.[35]

The walkout in the Motor City built on decades of organizing by teachers. The Detroit Federation of Teachers (DFT) emerged from the teacher unrest of the 1930s, seeking increased funding for schools in Detroit during the Depression, and, in the 1940s and '50s, became an integral part of what Jeffrey Mirel calls the "labor-liberal-black coalition" in the Motor City. Michigan had its own version of the Condon-Wadlin Act—called the Hutchinson Act and passed partially in response to a threatened teacher strike by Detroit teachers in 1947. Although the DFT had been able to influence education spending in the city even without collective bargaining, by 1963, Detroit teachers were paid less than most teachers in the surrounding suburbs. Catalyzed by the successes of the UFT and buttressed by the strength of labor in Detroit, the DFT threatened a strike in 1963 unless the board of education agreed to a representation election. The board agreed, and the Democratic Michigan legislature in 1965 amended the Hutchinson Act (a revision signed

into law by Republican governor George Romney); although strikes were
still banned, the new law outlined the collective-bargaining rights public
employees enjoyed in the state. In 1965, Detroit teachers leveraged a strike
threat to raise their salaries substantially, and two years later, the DFT followed through on its threat, shutting down schools for 300,000 children
for two weeks and ultimately gaining salary increases of $850 per teacher
for each year of the two-year contract.[36]

The most important strike in 1967, however, occurred in New York City.
The two-week strike, in which almost 50,000 teachers left their classrooms,
came just days after the State of New York passed a new public-employee
law to replace the draconian Condon-Wadlin Act. The new law—developed
by a committee headed by Wharton School of Business labor-relations expert
George Taylor—was designed to reflect a more enlightened labor policy in
New York. The need for a new law had been underscored in large part by
the two successful UFT strikes in 1960 and 1962, but the immediate impetus
was an illegal strike by New York City transportation workers in 1966.[37]
The new Public Employees Fair Employment Act, signed into law by liberal
Republican Nelson Rockefeller, inscribed the rights of teachers and other
public employees to collectively bargain; however, it retained the strike ban.
Although strikers would still be penalized, the penalties were less severe,
making it more palatable, in theory, for local authorities to punish them. If a
union went on strike, it could be fined $10,000 a day, courts could sentence
union leaders to jail time for violating an injunction, and the union could
lose its right to have employers automatically withhold dues from workers'
paychecks. At the same time, the law set up a state agency to mediate labor
disputes.[38]

When New York teachers struck in September 1967, they were the first
public-employee union to test the new law. The UFT called the action a
"mass resignation" in an attempt to avoid the penalties of the Taylor law,
but everyone clearly understood it as a strike. The strike was enjoined, the
union defied the injunction, and teachers negotiated a two-year contract
with an enormous increase in salaries and benefits. Though the union was
fined $150,000 and UFT president Albert Shanker spent fifteen days in jail
in December and January, it was clear that the new contract was worth the
price.[39]

Observers noted that New York law—in spite of the best intentions of the
legislators who authored it—simply could not stop strikes. As one report
after the walkout put it, "The teachers' work stoppage dashed the immediate hopes of those who saw the state's new Taylor law, hailed by Governor
Rockefeller as milestone legislation, as the solution to public employee

strikes." Further, a policy promoted as a "panacea" had foundered on the "basic problem" that "the public employee unions . . . feel they are justified in opposing a law they say will require them to bargain with their employer on an unequal basis . . . [and] that no penalties can force people to work when they do not want to."[40] Tom Wicker, in a *New York Times* editorial, perhaps best summed up the complicated view of teacher strikes across the nation in the fall of 1967. On the one hand, he blamed school boards and taxpayers for undervaluing teachers, arguing that "it is simply not possible" to "hand the public school teacher the vital responsibilities that American society now demands of them, and continue to pay them less than a truck driver, a football coach or a soda jerk." Wicker's comparison to male blue-collar workers who made more than teachers gives us some insight into both the class and gender politics of the era. Indeed, he asserted that teachers were as valuable, or even more valuable, than those who worked with their hands. On the other hand, Wicker also bemoaned the fact that teachers dealt with the problem through the same methods of "industrial unionism" that blue-collar unions might use. This development represented a veritable crisis the state could not manage: "This seems truly a season of disruption and disillusion. The war in Vietnam and the burning of the cities have exposed all our assumptions about ourselves to doubt and dismay. Now one of the most cherished of American traditions—the annual opening of free public schools for all—is being widely frustrated by teachers' strikes."[41]

There were fewer strikes in 1968, but those that occurred brought serious controversy. In January 1968, the nation's second-largest city—Chicago— narrowly avoided its first teacher strike because the board of education provided a new contract with $100 more per month for its 23,000 teachers.[42] At the beginning of February, AFT teachers in Cincinnati returned to work after a four-day strike. After closing almost half of the city's schools, the president of the Cincinnati Federation of Teachers was sentenced to five days in prison.[43] Then, at the end of the month, about half of Florida's 60,000 teachers walked off the job to protest poor school financing from the state legislature. Launched by the Florida Education Association (FEA), the action in Florida was in response to the fact that state lawmakers would not invest more money in improving conditions in the schools. Teachers provided en masse resignation forms (with the date left open) to local chapters at the beginning of the 1967–68 school year, and in February the chapters filed them simultaneously. It was clearly a strike, but because it was officially a resignation, was legal. This course of action, however, meant that teachers had no claim to their jobs, and district school boards had the option of replacing them. As Fred Hechinger argued, the Florida strike represented a "vital test" for the NEA. Indeed, the *New York Times* education reporter pointed out that the NEA had only just accepted the idea of supporting locals that went on strike and, known colloquially as a "company union" when compared to the AFT, had fallen behind as the smaller organization gained momentum in pushing aggressively for collective bargaining. The teachers certainly flexed their muscles; a third of the state's students were out of school as the strike entered its second week, and by the time the FEA called it off—after three weeks—in mid-March, over 18,000 teachers were reportedly still off the job. But the strike represented only a modest success; the legislature put more money in the schools but far less than teachers had argued was necessary.[44]

The Pittsburgh Teacher Strike of 1968

As New York City searched for a labor law that would work and teachers in Florida walked off the job in early 1968, Pittsburgh teachers went on strike to attain collective-bargaining rights and pay increases. The Pittsburgh Federation of Teachers (PFT), following the lead of big-city AFT affiliates that had won bargaining contracts in New York City, Detroit, and especially in Pennsylvania's largest city—Philadelphia—sought an election that would lead to exclusive representation for all city teachers.

As their counterparts did elsewhere, Pittsburgh teachers sought collective bargaining because they believed they were professionals who were not paid commensurately. As early as 1956—a year when the organization's total national membership was just over 50,000 and organizing funds were limited—AFT president Carl Megel identified the Pittsburgh area as one of a handful of places on which to spend scarce organizing resources. In early 1956, the AFT could only afford four full-time organizers and only because of a $1,500 monthly subsidy from the AFL-CIO; in a letter that June from Megel to the AFL-CIO's director of organization, he asked for a fifth full-time organizer in the Steel City to capitalize on the teachers' discontent. Six months later, Megel asserted to a regional organizing official that Pittsburgh represented the best organizing opportunity for the AFT in the entire nation.[45]

By 1968, as the strike began, that momentum led ordinary teachers to publicly articulate their frustration with salaries and working conditions. One teacher, in a letter to the *Pittsburgh Press*, pointed out that even though the city's teaching force, made up mostly of women, "often have had to go in debt for their initial educational expenses and must pay for their graduate work themselves," their pay lagged behind the average income of men with a high school diploma. Another teacher pointed to a recent study to argue that teachers typically worked fifty hours a week for a starting salary of $4,500 a year. He calculated that this worked out to $2.50 an hour (a small wage indeed considering Pennsylvania raised its minimum wage to $2.65 an hour in 1968).[46] A female elementary school teacher argued in a letter to the *Pittsburgh Post-Gazette* that "teachers have begged for crumbs for years—now we need to take action," while high school teacher Richard Price wrote, frustrated, to the *Pittsburgh Press* that "the teachers of Pennsylvania have been dumped in a corner and forgotten about by State legislators, citizens, and school boards. . . . To remain ignominiously dumped in such a corner is the most unprofessional position a teacher could take."[47]

The Pittsburgh Board of Public Education argued that the state's public-employee law—like New York's, passed in 1947 and forbidding strikes—precluded them from negotiating with "an exclusive bargaining agent." The PFT pointed to the Philadelphia contract and pressed the city to follow its eastern cousin's example. On February 29, 1,200 of the city's 3,000 teachers walked off the job and picketed city schools, causing over a dozen to close on the strike's first day.[48] In the midst of the strike, the Pennsylvania State Education Association (PSEA)—an affiliate of the AFT's rival—staged a one-day "professional development" walkout with teachers from across the state and marched on the capitol for a law requiring public employers to bargain collectively. As a result of the two actions, Governor Raymond Schafer—a moderate, establishment Republican in the Nelson Rockefeller mold—pledged a substantial package to raise teacher salaries across the state, and a local Pittsburgh judge agreed to supervise a representation election for city teachers with the understanding that state legislators would work to change Pennsylvania's public-employee law to allow exclusive representation.[49]

Like strikes elsewhere, the eleven-day Pittsburgh strike was steeped in controversy. Both of the city's two major daily newspapers—the *Pittsburgh Press* and the *Pittsburgh Post-Gazette*—criticized the strike for subverting the public interest. The *Press*, for example, surmised, on February 20, that "a strike by teachers is not likely to win favor even from a public composed largely of unionized workers—as is the case in Pittsburgh."[50] The editor's assumption about the city's "public" underscored the potentially problematic relationship of public-sector unions and the broader labor movement. On the one hand, the piece conceded that the city's private-sector unions—in a labor stronghold—might have enjoyed an automatic kinship with unionized teachers, but its further contention that "the public is getting sick and tired of being bullied by every union powerful enough to disrupt a public service" highlighted a possible rift between public- and private-sector workers.

A political cartoon in the *Post-Gazette* perhaps best shows the antistrike position of that daily newspaper. The paper's political cartoonist, Cy Hungerford, connected the threatened PFT strike to the larger national narrative of striking teacher unions. Referencing the Florida teachers' walkout, Hungerford's February 22 cartoon, titled "Contagious?," featured his stock representation of Pittsburgh—a corpulent combination of Benjamin Franklin and George Washington suited in an eighteenth-century style jacket, stockings, and tricorn hat—examining a sign reading "Florida Schools Quarantine—Teachers' Walkout Fever." In his hand, the anxious man, unsubtly named "PA PITT," holds a piece of paper suggesting that the Steel City is next: "SYMPTOMS OF FEVER AMONG PITTSBURGH TEACHERS."[51]

The February 26 edition highlighted the subversion of gender roles brought about by the teacher strike. Another Hungerford cartoon featured a dour female teacher presiding over a desk designated "Pittsburgh Federation of Teachers" and wielding a club labeled "Strike Threat." The school board, pictured as a boy, sings "'Readin' an' ritin' an' 'rithmetic'—sung to the tune of the hickory stick." On the floor lie the teacher's high-heeled shoes atop a piece of paper with the heading "Thursday Walkout."[52] The gendered dimension of the cartoon was integral to the message: Teaching was traditionally a female occupation (particularly in elementary schools) and the idea that a union gendered as female—even though its president, Albert Fondy, was, like most American union leaders, a white man—wielded considerable power over the humiliated little boy representing the school board was highly subversive. Indeed, the cartoon indicated how the inversion of power represented by teachers' collective organization could serve to upset gender roles at a time when a broader feminist movement even more consciously sought to do so.

Relatedly, a key feature of the discussion in Pittsburgh after the teachers struck on February 28 revolved around the illegal nature of the strike and highlighted the anxiety of Americans when even teachers could not be counted on to adhere to the law. Set directly below a cartoon titled "Inexcusable Absence" in the *Press*—in which two angry-faced striking teachers picketed in front of two perplexed and sad-looking white children, Judith Konnerth's letter to the newspaper asserted that "teachers decide when they will teach and when they will not. This is disobedience of the law. I am trying to teach my child to obey laws, to work toward changing a law instead of flagrant disobedience of laws established for the good of society."[53] Doris Worden connected a similar fear of declension to developments elsewhere: "The time has come for America to straighten up before we become a totally hopeless nation. I, for one, am sick and tired of strikes by garbage men in New York, hotel workers strikes that sound the death knell for Penn-Sheraton [there had been a recent strike at the Pittsburgh hotel], and now the teachers in Pittsburgh." To her, labor unions were greedy and irresponsible, and now that teachers were acting like other unionized workers, Americans no longer had "the right to recommend our way of life . . . as utopia for the rest of the world."[54] Taken together, these letters show that for many commentators, the stakes of the strike were much higher than just lost school days.

Another letter writer connected the teachers' behavior—that is, acting like other unionized workers—to a fear of the future of education in the city. On March 2, R. B. Russell believed that "once the teachers really submit to the rule of organized labor . . . the quality of the teaching will go down and down. A teacher who cannot think for himself is certainly not qualified to teach the young people of Pittsburgh."[55] Here Russell tapped into a long tradition of American mythology emphasizing individuality over collective organization, and which equated opportunity with the philosophical concept of negative liberty. Although this American usable past of settlers and frontiersmen goes back to at least the post–Civil War Wild West shows and Buffalo Bill dime novels, the 1950s and '60s came with the resurgence of the fear that big government, large corporations, affluence, unionization, and suburbanization had sapped Americans' ability to think. Radical sociologist C. Wright Mills, for instance, outlined a critique of an unchecked "power elite" at the top in conjunction with a labor movement led by the "new men of power," who had essentially, by becoming part of the system, ceased to be a viable opposition force. From a very different perspective, *Forbes* magazine editor William Whyte offered a dismal view of a corporate America in which the "social ethic" had subsumed the "Protestant ethic" and Yankee ingenuity was being eclipsed by mass conformity.[56]

In spite of attacks on the union's actions, the debate about the 1968 strike featured a good deal of support for higher teacher salaries, reflecting a widely held assumption that teachers, both as providers of opportunity and as stewards for the nation's youth, deserved to be remunerated fairly. A *Press* editorial from February 20, for example, which criticized the strike, nonetheless admitted that "there are many people who are convinced—particularly after learning how much garbage collectors in New York City earn—that teachers do deserve more favorable treatment."[57] Just as Tom Wicker in the *New York Times* had emphasized the distinction between blue-collar workers and teachers, so did the *Press* rely on disparaging manual labor in an argument to support higher salaries for white-collar teachers.

Another editorial, on March 4, asserted, "There is plenty of support both among voters and among officials, for the idea of tying teachers' salaries more in line with their professional standing." Ironically, however, although the piece advocated that teachers not strike or demonstrate at the state capitol (referring to the one-day demonstration by the PSEA in Harrisburg), it nevertheless admitted that teachers were "understandably" frustrated with years of unfulfilled promises of salary increases. In spite of the fact that Governor Shafer sought $25 million to raise teacher pay—a development pointed out by the editorial—as a direct result of the planned PSEA march and the teacher strike in Pittsburgh, the editor still argued that teachers should send a "well-informed and articulate small group to make their case in an atmosphere more suitable for discussion."[58] This sort of logic cut right to the heart of a basic contradiction in the postwar labor-liberal alliance: newspapers such as the *Press* and *Post-Gazette*—featuring liberal editorial pages supporting both higher salaries and the principle of collective action—failed to understand on a tactical level why it might be necessary for organized workers to withhold their labor.

The immediate outcome of the strike—in conjunction with the PSEA demonstration—brought salary increases for teachers across the state. By the end of March, the legislature and the governor reached an agreement on a $35 million pay raise to bring minimum teacher salaries in Pennsylvania to $6,000 a year (a 33 percent increase). The Republican governor defended the extra expenditure, arguing that "if the cost of government is going to increase through increased teacher salaries, then that cost will have to be paid."[59]

Not everyone agreed with the governor's logic, however. In a cartoon titled "The New Math," Hungerford portrayed the same blonde teacher from earlier cartoons sitting at a desk and pointing to a blackboard with "School Teacher's Pay Raise" chalked on it. Holding a ruler over the heads of two little boys named "Governor Shafer" and "State Legislature," she admonishes them with "That's Your Problem!"[60] If gender works primarily to signify power relationships, then the message of this cartoon was clear. The power structure in the state, signified as masculine, was symbolically turned upside down by aggressive collective action by teachers. Indeed, harking back to the critique of the "frenzied feminine finance" in Chicago in the early twentieth century, both the executive and legislative branches of the government, turned into little boys, had been emasculated by female teachers.

In spite of the gendered representations of teacher unions in 1968, many liberal politicians clearly believed teachers should be paid more in order to improve education in the state. Local school administrators, rather than fight to prevent pay raises and union representation, instead ratcheted up efforts to advocate for higher salaries in Harrisburg. In October 1970, for example, Pittsburgh schools superintendent Louis Kishkunas lobbied the Pennsylvania

General Assembly for new money for teachers. Kishkunas argued that this was necessary because in cities like Philadelphia and Pittsburgh, two-thirds of local revenues ended up paying for city services that benefited the entire metropolitan region, while suburban districts could use up to 70 percent of their tax dollars for education.[61] Although the proportion of spending on public education would not be equalized between city and suburb, both the City of Pittsburgh and the state contributed substantially more money to teacher salaries in the late 1960s and early 1970s than they had before the 1968 actions of the PFT and PSEA.[62] Militant action worked, and striking illegally represented an integral part of the effort.

Pennsylvania Act 195 and the Right to Strike

The commitment to pay raises from both state and local government in 1968 was significant. More important, however, the Pittsburgh strike led Governor Shafer—just two months after teachers returned to their classrooms—to establish a legislative commission to revise the state's public-employee relations law. Shafer undoubtedly responded to pressure from Democratic legislators—especially those from the Pittsburgh area like Leroy Irvis, who received political support from area labor unions—to change the law. But Shafer's action, including a reference to the Pittsburgh teacher strike as "a just appeal for a more democratic process" that unfortunately "has been made in violation of the law," also highlighted the reach of labor-liberal ideology.[63] In calling for the commission, he pointed to "a number of labor disputes by public employees with public employers where existing law has been disregarded, or in the opinion of some, has been held inadequate" in the "recent months."[64] Chaired by Leon Hickman, a retired corporate executive, the commission's findings evidenced the conflicted nature of public-sector labor rights in the late 1960s. While the commission, like legislators in other states, tried to deal with the problem by treating public workers the way private-sector workers had been treated by the Wagner Act, the emphasis on preventing strikes made the new law problematic.

After holding public hearings in Philadelphia, Pittsburgh, and Erie, and conferring with public officials from Wisconsin, Michigan, and New York—all states with nationally recognized public-sector labor laws—the commission, in June 1968, called for a new statute to recognize "the right of all public employees to bargain collectively." This part of the recommendation did not depart from laws in similar states; it differed, however, in conferring a "limited right to strike" on most public employees in Pennsylvania, including public school teachers. Michigan's law had made it more difficult than in other states to punish strikers, but no other state allowed public employees to strike.[65]

Surprisingly, by ensuring that public employees enjoyed the right to strike, the commission hoped not to incite more strikes but instead to prevent them. Under the recommendations, public employees could only legally strike after all efforts to solve a labor dispute through bargaining—mediation, fact-finding panels, and arbitration proceedings through the state department of labor—failed. Only then—presumably when a school board or other local government was being unreasonable—could a strike occur. The commission report argued that a dispute would only lead to a strike when the public supported it because "strikes can only be effective so long as they have public support. In short, we look upon the limited right to strike as a safety valve that will in fact prevent strikes."[66]

In advocating this course of action, the commission intended to avoid the criticisms public unions levied on policies like the Taylor law, which gave public employees the ability to bargain but still gave the state a major advantage in the bargaining relationship unless unions broke the law. But the Hickman commission did not go all the way: if the "safety valve" represented by the right to strike failed to prevent a walkout, the commission recommended that a strike could be enjoined by local officials if the action violated the "welfare, health, and safety of the general public." The courts would be charged with defining such a violation, and the commission recommended severe penalties—substantial fines and even jail time for union leaders—if strikers refused to honor an injunction at that point.[67]

In conferring the right to strike, the proposal evidenced two key assumptions of labor liberalism in the late 1960s. First, the commission highlighted the assumption—as was the case in the construction of the Taylor law in New York—that rational administration of government policy could solve conflict and ensure government services. This assumption stemmed logically from the major rationale for the federal government's interest in supporting labor unions in the 1930s: ensuring the flow of interstate commerce by giving workers the right to unionize. Second, the commission highlighted the view that unionization on the part of workers was necessary to counteract the excessive power of the employer. Prominent postwar liberal intellectuals like John Kenneth Galbraith and Arthur Schlesinger Jr. clearly evidenced the assumptions of the Wagner Act in this regard in the postwar era, arguing that labor unions in the private sector served as crucial checks on the power of corporations.[68] The commission's recommendation took a step in promoting the counterbalance of state power represented by public-sector workers, but as we shall see, the injunction provision would continue to force teachers to strike to balance out this asymmetry.

The right-to-strike provision proved controversial among Republican legislators in the state's upper chamber, but another threat of a teacher strike in Pittsburgh in 1969 and contract negotiations in Philadelphia on the horizon pushed the Pennsylvania General Assembly in 1970 to pass the new public-employee legislation.[69] Titled the Public Employee Relations Act (or PA 195), it was virtually identical to the Hickman commission proposal and represented a far-reaching experiment in public-employee labor relations. Indeed, Pennsylvania became the first industrial state in the United States to formally permit public-employee strikes in any circumstances.[70]

Public-sector unions—especially teacher unions—were pleased. PFT president Albert Fondy called the legislation "a tremendous advancement" and believed that "Pennsylvania . . . now becomes one of the most progressive states in the nation."[71] It was not only unions that touted the law, however. The Pennsylvania School Boards Association—the umbrella organization representing Fondy's adversary in 1968—also "hailed" the new law.[72] And finally, the Republican Shafer, who expressed misgivings about public-sector workers' right to strike, nonetheless signed the law—a crucial point given the Republican-held state senate's inability to override a veto. Shafer's rationale lay well within the liberal calculation that labor conflict could be mitigated through laws giving public-sector workers an attenuated version of the labor rights workers enjoyed in the private sector. The governor believed that "without such a law, we will have chaos in our public institutions." Buttressed by responsible use on the part of both employers and employees, he asserted, the legislation allowed "differences" to be "resolved at the negotiation table, not on the street in a picket line." Shafer concluded that "we today step into a new era of labor relations."[73]

Even media outlets that had opposed public-employee strikes expressed optimism about the ability of PA 195 to provide an orderly system of labor relations. The *Pittsburgh Post-Gazette*, which had been so critical of the teacher strike in 1968, averred that the law "recognizes the facts of our complex industrial society" and believed that it might be able solve the problem.

> All we know is that the Legislature, with an assist from a commission of public-spirited citizens, sought to deal fairly and effectively with a sticky public issue. . . . But, of course, there is no iron-clad guarantee that the law will work. New York State's unhappy attempts to legislate in the field of public employee-employer relationships are not reassuring. Perhaps there is no way to stem the

spreading militancy of public employee organizations. But the effort has to be made if a viable society is to be preserved.[74]

Perhaps the most sanguine editorial came from the Harrisburg daily the *Patriot*. The paper pointed out that "astute observers believe [PA 195] may become the most significant law to emerge" from the state legislature "in a generation." The editorial presciently argued that the law would inspire a surge of labor organizing, as well as necessitate both an increase in the responsibilities of the state labor relations board and a "vitally significant" role for the local court system in defining the scope of the law. Although much still needed to be done, the editor believed that, with proper implementation, it was possible that "we have witnessed the birth of a wondrous new thing"—a policy that guaranteed labor rights for public employees while also preventing strikes.[75]

PA 195 in Action: Philadelphia and Pittsburgh

The success of the new law would be determined by the labor conflict in Pennsylvania that followed, and observers elsewhere watched to see if it would actually prevent labor stoppages. Indeed, in the early 1970s, discussion of PA 195—and the efficacy of liberal-labor policy in the public sector—played a prominent role in the public discussions of teacher strikes in both Philadelphia and Pittsburgh.

By the time PA 195 became law in 1970, there was much more teacher labor conflict across the country. In September 1968, in what had become an annual rite, the school year began with labor stoppages. The most nationally prominent was in New York, where 54,000 New York City schoolteachers walked out on September 9 (a development treated in more depth in chapter 2). Teachers walked out again in East St. Louis, Illinois, while across the Mississippi River, St. Louis, Missouri, only averted a strike when teachers for AFT Local 420 (the St. Louis Teachers Union) voted by a narrow margin to obey an antistrike injunction. Teachers also struck, among other places, across Michigan; in East Haven, Connecticut; and at West Homestead, Pennsylvania.[76]

In May 1969, Chicago experienced its first strike; teachers there sought sizable pay increases and better teaching conditions. On December 1, 1969, Denver, Colorado, teachers affiliated with the NEA called off their victorious two-week strike after winning substantial salary improvements and amnesty from legal action for striking. Providence, Rhode Island, teachers also went on strike for eleven days. In March, Jersey City teachers shut down schools for four weeks; although they struck illegally and faced an injunction, the school board agreed to increases of $1,700 for starting salaries over two years. In May 1970, Boston teachers walked out of classrooms, and the union's president was sentenced to thirty days in jail.[77]

The largest strike that year, however, was in Los Angeles, the second-largest school district in the United States. Beginning in mid-April and lasting for four weeks, teachers in Los Angeles went on strike for higher salaries and smaller class sizes. The state's public-employee law—passed in 1965—gave teacher organizations the right to "meet and confer" with school boards but barred exclusive representation or collectively bargained contracts. Without these rights, the NEA and AFT affiliates merged in 1969 in order to pool resources and push for a de facto bargaining relationship. By the time the strike began, the new United Teachers of Los Angeles (UTLA) had 13,000 members. On April 13, over half of the city's 25,000 teachers stayed off the job, and many students walked out in solidarity with their teachers. The district won a restraining order, and while most teachers ignored it, it did slightly reduce the effectiveness of the strike; the 60 percent of teachers out on the first day dropped to just 50 percent by the second week of the strike, and most schools—although the quality of education for virtually all of the city's 700,000 students almost certainly suffered—remained open. The strike ended when the board offered a 7 percent wage increase, which was contingent on the passage of a state ballot referendum—Proposition 8—to raise more revenue for the schools. The teachers, by a three-to-one margin, voted to accept the deal and offered to give up the possible raise in exchange for funding remedial programs for students.[78]

Several aspects of the public discussion of the Los Angeles strike are instructive. First, with teachers in the city split almost down the middle on whether to strike, teachers' rationale for their individual decisions highlights the tension endemic to teacher militancy in the early 1970s, particularly in Los Angeles, where teachers were not as quick to build power as their eastern counterparts. Eleanor Reicher, a reading specialist in an elementary school with average class sizes of thirty-eight that was located "on the fringes of a ghetto area," found, as a "professional," the possibility of a strike "abhorrent." She supported it, however, because she believed "the people of the city and state need a dramatic demonstration of just how bad things are." She asked the public to "imagine yourself working until late at night to prepare lessons and having them fall flat because there are too many children in your classroom and many of them aren't prepared for the experience of coming to school." James Baxter, however, chose not to strike and viewed professionalism differently. He believed that "a strike is a confession that we teachers have lost our unique and individual identity and purpose and have become stiff-necked, petulant, and defensive—problem-makers, not problem-solvers."[79]

Second, perhaps the most important public comment during the conflict came from Ronald Reagan. Reagan, who had become a star of the GOP's conservative wing after his national convention speech in 1964, won election as governor of California two years later in large part because of his campaign promise to shut down student protest at Berkeley.[80] As a staunch opponent of big government, bureaucracy, and tax increases, Reagan opposed Proposition 8, which helped lead to its defeat. Most interesting about Reagan's view of the Los Angeles teacher strike, however, is that his famous stand against the Professional Air Traffic Controllers Organization (PATCO) in 1981 seems to have found a good deal of its inspiration from the Los Angeles teacher strike.[81]

In Reagan's view, private- and public-sector unions were distinct, and he made this view apparent in August 1981, when, as president, he summarily fired over 10,000 striking air traffic controllers. Reagan argued:

> Let me make one thing plain. I respect the right of workers in the private sector to strike. Indeed, as president of my own union, I led the first strike ever called by that union. I guess I'm maybe the first one to ever hold this office who is a lifetime member of an AFL-CIO union. But we cannot compare labor-management relations in the private sector with government. Government cannot close down the assembly line. It has to provide without interruption the protective services which are government's reason for being.[82]

In 1970, Governor Reagan's view of the teacher strike clearly anticipated his response to PATCO, even down to the language he used in criticizing teachers. "In private industry," the future president argued, "the strike is a legitimate tool which can be used to close down a business until agreement has been reached through negotiations. In government, we cannot shut down the assembly line. Government is constitutionally and morally bound to provide certain basic services. Government has no choice but to continue operations any way it can." If Reagan's view of the strike sounded a good deal like Roosevelt's remarks on public-sector strikes in 1937, the larger significance of his pronouncement stemmed from the fact that driving a wedge between public and private workers by emphasizing the qualitatively different form of employment—as we shall see in his public critique of the New York City fiscal crisis in 1975—would become a staple of Reagan's shrewdly calculated political strategies.[83]

By the beginning of the 1970–71 school year, as Fred Hechinger explained, many teacher labor disputes had begun to pivot on a chronic "money squeeze." Hechinger pointed out that school boards faced budget difficulties in part because of higher teacher salaries that had been collectively bargained. Many school districts spent down cash reserves and, seeking new sources of revenue in taxes, had started to run into stiff taxpayer resistance. The economic recession beginning in the fall of 1969 exacerbated the problem.[84] Clearly, the Los Angeles strike hinged on school finances—state aid during the Reagan administration for Los Angeles had not kept up with its expenses; one estimate in 1970 pointed out that California only provided 28 percent of the budget for Los Angeles schools, whereas it provided 36 percent on average for other school districts[85]—but Hechinger cited localities in Ohio, Oregon, Missouri, and Pennsylvania as particularly in trouble that fall; of these, Philadelphia represented the most serious. In September 1970, just a month before PA 195 was to go into effect, the city's fiscal problems led teachers in Philadelphia to strike.

A vanguard teacher union in a city with a robust history of organized labor, the Philadelphia Federation of Teachers (PFT) signed their first union contract in 1965. Teachers enjoyed higher salaries and smaller class sizes in the years that followed. When teachers began negotiating a new contract in 1970, however, they faced inauspicious circumstances. First, Pennsylvania standardized the instructional day for all of the state's school students—to five and a half hours a day—in 1968, and Philadelphia high schools only provided four and a half. The board of education needed to rectify the imbalance in the new contract. Second, Philly schools faced a $50 million budget shortfall. Although the fiscal shortfall in Philadelphia would get much worse, the scale of the deficit was unprecedented in 1970. The president of the Philadelphia Board of Education at the time called the situation "dreadful," and a reporter for the *Philadelphia Inquirer* had pointed out the previous February that "Philadelphia is on the threshold of becoming the first big city in America to close its public schools. . . . It is not hyperbole to say the public school system is on the verge of collapse."[86]

Indeed, Philadelphia shared the trajectory of many industrial cities of the Northeast and the Midwest after World War II. As other groups had before, unskilled African Americans migrated in large numbers to the City of Brotherly Love in the 1940s and '50s. During this era, however, many whites left the city to go to suburban developments, abetted by the federal government through housing policy and the federal highway system. Also, many industries shifted production from Philadelphia to outlying suburban areas (such as Bucks County, in the northeastern suburbs) or to states in the South and West. Indeed, in the 1950s and '60s, Philly lost almost 100,000 industrial jobs, particularly in the textile industry, where unskilled African

Americans migrating to the city might have found steady, entry-level employment. With limited economic opportunities in the city and wealthier Philadelphians moving to the suburbs, property-tax revenues could not keep up with the services necessary to run an education system in a city like Philadelphia, particularly as teachers put pressure on the school board for higher salaries and benefits. Local school financing in the United States and businesses' search for lower taxes and lower wages in the postwar years alone would have endangered the schools' financial situation, but the federal government's interventions in the housing market altered the political economic structure of the city almost overnight. Furthermore, from 1950 to 1970, the demographics of Philadelphia changed immensely: the white population decreased while the black population almost doubled, and the city's racial makeup shifted from 82 percent white in 1950 to about 65 percent in 1970. African Americans became increasingly concentrated in specific neighborhoods like North Philadelphia and West Philadelphia, and the white working-class families who remained in the city lived mostly in Northeastern Philadelphia or South Philly.[87]

Teacher demands for higher salaries put additional pressure on the school district, but compared to other major cities, these salaries were hardly excessive: while Philadelphia teachers' salaries increased substantially in the first few years of collective bargaining, starting pay for teachers in Philly was lower than in comparable cities like Chicago, Detroit, and Washington, DC, and just a bit higher than in Baltimore and Cleveland.[88]

In spite of the prima facie evidence that higher teacher salaries caused the schools' fiscal problems, much of the public commentary on the budget shortfall focused on the need for the state government to contribute more money to the city school system. An *Inquirer* editorial on February 9 asserted that the formula for state aid was "rigged." Philadelphia, the piece argued, received only about "one-third of its school funds from the state while other school districts in Pennsylvania are getting as high as 80 percent of their budgets paid for by the state." The editorial proposed that the city council, school board, and mayor's office work together for "a fair shake from Harrisburg."[89]

Philly teachers, mindful of the city's budget problems, did not strike that September until pushed into it. By then, the board and the PFT had reached an impasse: They were $25 million apart on salaries, and teachers wanted the board to add new positions to lower class sizes. In contrast, the board wanted teachers to spend more of their day teaching instead of in preparation, or, as the media often called it, "free time."[90] The board closed the schools on September 10, and the teachers promptly went on strike. The schools reopened four days later when the board and the union agreed to negotiate for thirty days.[91] That deadline passed, and the teachers walked out for three more days. Many schools closed and about half of the city's 231,000 students stayed home. The teachers went back to work after reaching an agreement with the school board for a fairly substantial salary increase as a lump sum divided by the city's teachers—estimated at about $900 per teacher each year—and twenty minutes more instruction each day to ensure the five-and-a-half-hour requirement.[92]

Strike negotiations, occurring just after PA 195 became law, took place under a media microscope. Although it did not actually go into effect until October, politicians, political commentators, and community activists alike called on both sides to accept the "spirit" of the new law to avert the consequences of the labor stoppage. In a letter to both Frank Sullivan, president of the PFT, and Richardson Dilworth, president of the board of education, Mayor James Tate "strongly urge[d]" both sides to "institute the impasse procedures contained in the New Public Employee Relations Act without delay." Tate, indeed, was optimistic that if both sides accepted the "intent" of the law, they could avoid a prolonged strike. PA 195, he believed, represented the "joint product of collective bargaining experts for both labor and management that designed into law the best techniques and procedures known to facilitate agreement."[93] On August 28, 1970, an editorial from local CBS affiliate WCAU-TV asked Governor Shafer to "call both sides to Harrisburg . . . and convince the Teachers Federation to accept the intent of the law even though it is not yet legally binding."[94] Mrs. Gail Aronson, secretary of the Loesche Public School Home and School Association (HSA)—in Philadelphia, the HSA equated to the more widely known Parent-Teacher Association (PTA)—called on the PFT to "accept mediation, arbitration, [or] fact-finding."[95] Though not completely optimistic about the effectiveness of the law, Aronson's letter evidenced the widespread belief at the time that well-thought-out policy could avert labor stoppages.

Although no one desired a shutdown of city schools, it should be noted that many argued that the teachers had a legitimate reason for striking. Indeed, Philadelphians clearly believed education was important and that well-paid teachers represented a key component of quality education. An editorial in the *Evening Bulletin* on the day before the September school closing criticized the union, arguing that in "almost every respect [the strike] is wrong." Nonetheless, the piece admitted that "the public . . . is not unsympathetic to the teachers' need for economic equity. Considering educational requirements and their importance to society, they should not have to moonlight on second jobs to achieve a decent standard of living."[96] In a letter to the city's other major newspaper, the *Inquirer*, Ralph Smith contrasted the largesse of city spending in order to host the 1976 bicentennial celebration with its inability to pay decent teachers' salaries: "More than a billion dollars may be spent. . . . It is hard to believe the Federal Government will pay for all this, which means the Philadelphians will again be hit hard. And we can't even get 24 million dollars together to send our children to school!"[97]

Still, though many interlocutors during the Philadelphia strike of 1970 believed the strikers' cause was just, others evidenced a growing sense that union tactics signaled a larger "crisis." Political commentators and community activists pointed to a drastic decline in respect for "law and order" and worried about the fate of a nation in which teachers could "defy" the law. Indeed, some of these observers, like Governor Reagan in Los Angeles, viewed public employees as distinct from private-sector workers. On September 12, for example, Alexander Layman, a "tax-paying Philadelphia business man," wrote to PFT president Sullivan. Layman did "not think that your organization as well as the Board of Education should conduct negotiations in the same manner as the teamsters or automotive workers." Focusing on teachers' roles as caretakers, he went on to ask the union to "use your good common sense to get the kids back to school and then arbitrate while they are learning."[98]

Mr. and Mrs. Irv Forman, parents from the white working-class neighborhood of Northeastern Philadelphia, also asked the teachers to deviate from the methods of unions in the private sector. Although the Formans "sympathize[d] with your situation, we feel that you are wrong if you will not work during mediation. You seem to have forgotten that you are teaching children not machines." The stakes of the strike, however, were even greater than a few days of lost classroom time. Indeed, the Formans argued that, by striking, "you will only drive the middle class citizen out of the city to the suburbs where you don't have to worry about school strikes."[99] As we shall see in the much longer strike of 1972–73 (see chapter 3), the Formans represented a growing constituency of white working-class Philadelphians who viewed the "middle-class" citizen through a racialized lens and worried that the city would lose its "productive" core.

The president of the HSA for Fitzpatrick High School—also located in Northeastern Philadelphia—asked Sullivan to consider the example teachers set. Invoking the rhetoric of "law-and-order" politicians like Alabama Democrat George Wallace, Republican president Richard Nixon, and Philadelphia's own Democrat police chief Frank Rizzo, Dorothy Ravich believed the meaning of the strike ranged well beyond salaries and budget deficits: "In these difficult days of campus disorders and violent demonstrations, we want the children in our school to see an example of adults settling their differences in an orderly manner within the laws of our society."[100]

If community activists like Ravich sounded the alarm of declining law and order, local media pushed even further after the teachers violated a court injunction during the second strike. On October 19, for instance, the *Inquirer* placed the walkout in the context of the new public-employee law, set to go into effect two days later. The newspaper's editor doubted whether the "teachers and their leaders who have refused to obey the no-strike law presently in force will show any more respect for the new statute when it takes effect." The *Inquirer* believed the teachers' actions threatened the very future of the city because their disrespect for the law "may go a long way toward explaining the academic and disciplinary failure of our public school system."[101]

Not only did the illegal nature of the 1970 strike raise questions about PA 195 and deepen observers' sense of crisis, but it also represented just one part of a series of strikes that fall, many of which featured teachers who broke the law. In September, for example, local authorities in Connecticut jailed fourteen leaders of a 2,000-strong strike by New Haven and West Haven teachers (both AFT locals); two months later, between 500 and 800 (accounts varied) Hartford AFT teachers, out of a total force of about 1,800, went out for three weeks, returning to work without wage gains or even freedom from disciplinary action when the strike was enjoined. Clearly, the lack of solidarity—teachers from the local NEA affiliate declined to join—had been instrumental in defeating the strike.[102]

In early 1971, another strike in Pittsburgh again ignited a local debate about the effectiveness of Public Employee Act 195. Negotiating their second contract, Steel City teachers walked out in 1971 over fringe benefits, working conditions such as class size and special-education programs, and, most important, salary increases. Many AFT affiliates in the late 1960s and '70s sought to lower class sizes both for pedagogical effectiveness and to make teaching less onerous, and the PFT effort was part of that wave. Pittsburgh teachers also sought to bring their salaries into line with urban school districts like their cross-state brethren in Philadelphia. Although the school board and the Pittsburgh Federation of Teachers disagreed about the number of strikers, at least 60 percent and perhaps as much as 90 percent of the city's 3,200 teachers stayed off the job the entire week of January 4, 1971.[103]

Though Pittsburgh teachers now had the right to strike under PA 195, school districts could slow down the process by seeking various forms of mediation. On the second day of the strike, school superintendent Louis

Kishkunas asked the state labor relations board to establish a fact-finding panel and ordered the teachers back to work. PFT president Albert Fondy, however, argued that the teachers were not obligated to work without a new contract. The school board, on the strike's fourth day, sought an injunction to compel the teachers to return, but the board and PFT reached a contract agreement over the weekend and the injunction proceedings were moot. The new agreement gave teachers generous salary increases, ranging from $1,800 over two years for beginning teachers to $3,500 for the most experienced. The school board also agreed to hire eighty-six new teachers to reduce class sizes. This represented a clear victory for teachers, although the new starting salary of $7,800 a year still lagged behind the starting salary of Philly teachers following the contract they signed in 1970.[104]

It is difficult to assess public opinion during the 1971 strike for two reasons. First, the strike occurred in the midst of a two-week-long Pressmen's Union strike that shut down both newspapers for the entire duration of the teachers' walkout. Thus, by the time the newspapers began printing again, the strike had ended, and public discussion was much less relevant. Second, the strike was short and did not reach the point at which striking teachers broke the law.

Nevertheless, it is clear from the historical record that local opinion makers believed that the strike cast further doubt on whether Public Employee Act 195 would prevent public-sector labor strife. For example, on January 18, 1971, in the first issue off the presses following the newspaper strike, a *Pittsburgh Press* editorial argued that the teacher strike's real winners were the teachers, who gained large pay increases, while the parents received only the "assurance of a two-year peace from the PFT," which was really no win at all, since the PFT "has been raising strike threats with disturbing frequency in the past three years."[105] The real problem, however, was that the union failed to abide by the spirit of the law. The piece asserted that the PFT would "have none of" the intermediate steps—fact-finding, mediation, and so on—that could prevent strikes; in fact, the author argued—in bold print—that the PFT "left no doubt that it wanted to show its muscle in the most forceful way it could." It wondered if the new law would ever "become a useful tool for preventing teachers' strikes."[106]

For the *Post-Gazette*, this question was of the utmost urgency. Connecting the teacher strike to the newspaper strike, the featured editorial on January 18 argued that "it is a tragic commentary upon our social structure that a willful few can exercise monopoly power to inflict grave injury upon the public." And, in spite of new legislation to prevent public-sector strikes, the teachers "flouted" the law anyway. The editorial concluded that "irresponsibility is becoming the hallmark of too many people entrusted with the public welfare."[107] Even the televised national news reported on Pittsburgh in January 1971; NBC, for instance, featured several reports from David Brinkley on the strike, noting that it served as a test case for whether PA 195 could successfully stanch public-sector labor conflict.[108]

Conclusion

This chapter charted the development of teacher unionization during the course of the twentieth century and showed the explosion of teacher militancy in the 1960s. By the early 1970s, teacher strikes represented prominent arenas of debate over the political role of public-sector labor unions and the effectiveness of the state in dealing with the problem. Divisive walkouts in Pennsylvania led politicians, commentators in the media, and ordinary citizens to begin questioning the efficacy of the state as well as the legitimacy of teachers unionizing as other workers did. In the next chapter, we will see how these developments became enmeshed with another key political problem: rectifying the structures of racial inequality in many of the nation's largest cities.

CHAPTER 2

Teacher Power, Black Power, and the Fracturing of Labor Liberalism

In April 1968—just ten days after the assassination of Martin Luther King Jr.—political scientist Charles Hamilton published a lengthy piece on Black Power in the *New York Times*. The coauthor, with Kwame Ture, of the most influential explanation of Black Power in the United States, Hamilton wrote to define the concept and to explain to the nation why it was necessary for racial equality. Further, for white America, he pointed out why community power was necessary to quell the insurrectionary violence in the cities. Hamilton argued that Black Power should not be "equated with calculated acts of violence" because such action alone would not bring equality. At the same time, he asserted that African Americans—in Los Angeles, Newark, Detroit, and elsewhere—were "legitimately fed up with intolerable conditions." The main concern of Black Power, according to Hamilton, was "organizing the rage of black people and . . . putting new, hard questions and demands to white America." In particular, the Roosevelt University political scientist urged three goals: mitigating the "growing alienation" of African Americans and their "distrust of the institutions of this society"; creating "new values" and a "new sense of community and of belonging"; and establishing "legitimate new institutions" in which black people—largely excluded previously—might actively participate.[1]

A key aspect of Hamilton's vision of Black Power was the inner-city education system. Systematic desegregation of schools appeared to be impossible—at least for the "foreseeable future." Indeed, in Ture and Hamilton's larger argument for Black Power published in 1967, they focused prominently on white resistance to integrating Intermediate School 201 in Harlem that year in order to show that community control of the school represented the only practical solution.[2] Beyond shifting responsibility for education policy, Hamilton's piece in the *Times* also argued that schools should become the "focal point of the community": "School would cease to be a 9-to-3, September to June, time-off-for-good-behavior institution. It would involve education and training for the entire family—all year round, day and evening. Black parents would be intimately involved as students, decision-makers, teachers." Not only did Hamilton want to route welfare funds through schools in order to eliminate the stigma of caseworkers investigating families for eligibility, but he wanted the teaching force to include both "professionals" and parents who could teach the skills they already knew. Indeed, the school would "belong to the community" and bring together "children, parents, teachers, social workers, psychologists, urban planners, doctors, [and] community organizers."[3]

As Hamilton outlined this expansive vision for urban education, Black Power activists were already trying to make it a reality. In Philadelphia, students and other community activists had by 1968 forcefully pressed demands for more black teachers and principals and for the teaching of black history in the classrooms. Buttressed by support from liberal Republican mayor John Lindsay and with outside funding from the Ford Foundation, a community-led school board in Brooklyn was working to reshape the district's personnel to better reflect the goals of empowering black youth. In Newark, Black Arts poet Amiri Baraka worked with other activists to develop black electoral and community power through the Committee for a Unified NewArk (CFUN); after this effort bore fruit, CFUN tried to change the parameters under which the city's majority-white teaching force worked.

These actions led to a series of dramatic conflicts with unionized teachers. During the course of the 1960s and '70s, teachers fought tooth and nail against school boards reluctant to collectively bargain. They went on strike at great risk, almost always breaking the law in the process, to gain higher salaries and more control over where they worked and under what conditions. For some teachers, this new power meant avoiding teaching students whom they viewed as dangerous and difficult to teach.[4] For some white teachers committed to improving education in black schools, many assumed that only through teaching middle-class, individualist values could blacks overcome the "culture of poverty" that entrapped them.[5] For other teachers, however, increased teacher power clashed with the demands by Black Power activists that teachers should shoulder more caretaker responsibilities in the schools. The notion that teachers—in elementary schools, virtually all women—should provide extra care to black children to compensate for the institutional failures of city school systems was in tension with a view of professionalism that meant not being saddled with the seemingly endless demands for the care of children by male principals as had often happened in the era before collective bargaining.

The more "professional" work environment and an entrepôt into the middle class many urban teachers now enjoyed had only come through great struggle so recent that teachers remembered it well. Militant teacher resistance thus clashed with the efforts of Black Power advocates to wrest greater control over teachers' working conditions. Strikes by teachers over challenges in New York City and Newark represented the most spectacular manifestations, but conflict between majority-white teacher unions and Black Power advocates erupted in Chicago, Philadelphia, and Detroit as well.

This chapter charts those developments. Much scholarly work has focused on the conflict between unionized teachers and African American community control in New York City, exploding in the Ocean Hill–Brownsville crisis that led to three strikes by the UFT in 1968. I argue here that we should also see how the unsettled status of public-employee labor rights exacerbated that conflict. Not nearly as much work has focused on two strikes in Newark in 1970 and 1971 that, although occurring in a smaller city, were equally critical. I show that the struggle there caused a national controversy as profound as that of Ocean Hill–Brownsville.

In both of these cases—as well as in conflict elsewhere—unionized teachers emerged mostly victorious. The African American community—a bloc of growing political power—in both Newark and New York City, however, would be largely alienated from the teacher unions. When cities like New York and Philadelphia faced dramatic showdowns over fiscal crisis in the 1970s and early 1980s, groups that might otherwise have shared a common interest in the future of the school system had been driven apart. Some of the white working and middle classes in these cities, for a time, backed union teachers as proxies for their opposition to Black Power. Because teachers' strikes were illegal and teachers had also begun to serve as representations of a state unable to help private-sector workers during a time of inflation and economic downturn, however, critics began to link the black poor and unionized teachers as illegitimate takers of productive resources. I conclude this chapter by examining discourses around incarceration for illegal teacher strikes. I show how the incarceration of female teachers violated notions of gender propriety and threatened the American family, and, using a series of letters sent to the president of the AFT—incarcerated for violating an injunction in 1970—I highlight the emergence of a newly imagined political realignment in which critics increasingly characterized urban public school systems as needless wastes of tax dollars earmarked for those who refused to obey the law.

Crisis in Ocean Hill–Brownsville

The Ocean Hill–Brownsville conflict in New York City in 1968 has rightly garnered its historical reputation as the most spectacular conflict between unionized teachers and black nationalists over control of urban public schools in the late 1960s. On May 9, 1968, the governing board of the experimental Ocean Hill–Brownsville school district terminated fourteen teachers, sending each a letter to report to the central board of education for reassignment. The action represented the confluence of two major developments. First, white resistance to integration in New York and the board of education's intransigence in listening to African Americans' concerns about the decrepit school buildings and overcrowded classrooms in the city led activists to seek "community control" of overwhelmingly black schools in some of the city's poorest neighborhoods in Harlem and Brooklyn. Second, white liberals—represented by the city's Republican mayor John Lindsay and the philanthropic Ford Foundation—feared the urban conflagrations that had occurred each summer since the Watts uprising in 1965 and viewed black control of school policy as a way to mitigate the problem of racial inequality.[6]

In 1967, the New York City Board of Education offered community control over education on an experimental basis to the schools in the Ocean Hill–Brownsville section of central Brooklyn, where 95 percent of the students were nonwhite and the students' test scores across the district were well below the average in the city.[7] The Ford Foundation provided grant funding to the newly elected school board, and, by May 1968, the board—consisting of community parents and acting on the recommendation of its African American "unit administrator" Rhody McCoy—sought more power over personnel. McCoy wanted the teachers transferred because he believed they did not support the community-control project and wanted to replace them with teachers—white or black—he believed were committed to educating black students.

After the termination notices, UFT teachers in Ocean Hill–Brownsville walked out for the rest of the school year. The issue was due-process rights. According to their collectively negotiated contract, teachers had the right to a hearing when their employment was terminated, and the notice from the board had "terminated" their employment. The UFT argued—even though the notice also asserted that they were to be reassigned—that teachers had the right to a hearing before such an involuntary transfer. Beginning the following September, Shanker led teachers across the city—54,000 in total at the beginning—in a series of strikes that shut down schools across New York

City. The first strike lasted two days, and the board of education ordered the teachers back into the classrooms; resistance from the Ocean Hill–Brownsville district administration sent the teachers back on strike for two weeks. Again the teachers returned to the schools, but following violent clashes between union teachers and community-control advocates, the UFT went on strike again, this time for five weeks. Shanker called for the removal of McCoy and the local school board and the dismantling of the local school district; while he did not get either, the strike ended in mid-November when the state Department of Education stepped in to ensure the teachers would return to classroom assignments and, going forward, that all teachers would have the right to a hearing before an involuntary transfer.[8]

The crisis in Ocean Hill–Brownsville has received ample scholarly treatment. Jerald Podair's analysis of the strikes remains the definitive account, and he argues persuasively that the conflict worked to realign the political parameters of the city. A tenuous liberal coalition of blacks, Jews, and Catholics unraveled during the course of the crisis, and by the end, white outer-borough New Yorkers saw their interests as very different from the African Americans living in the poorest neighborhoods of the inner city. The breakdown, according to Podair, stemmed from a division between "two New Yorks" that had emerged after World War II. One New York—that of whites—believed in liberal, individualist pluralism. These whites—many of whom came from immigrant families—became successful in the city's growing public sector. They believed that anyone else who worked hard could do the same. The other New York—that of working-class African Americans and Puerto Ricans—faced institutional barriers: lack of access to good housing, limited employment opportunities, and inferior schools. They viewed the school system as inherently unfair, and, in Podair's words, "emphasized mutuality and cooperation; the cultural legitimacy of the black poor; the use of the cultural resources of the black community as a form of currency in the local and national marketplaces; and a pluralism based on community and group distinctiveness."[9]

Indeed, the striking teachers and community-control activists viewed the key stakes of the conflict through totally different eyes. Other important research on the conflict deepens this point. As Daniel Perlstein argues in his account of Ocean Hill–Brownsville, "at precisely the moment when black parents were challenging school officials' failure to combat racial inequality, the UFT argued that teacher professionalism precluded parents from exercising significant authority in the schools." Perlstein's careful treatment of the many different views on the conflict highlight just why the strike was so important. Like Podair, he explains the UFT's commitment to "race-blind" merit, and he also explains the limits of the integrationism of civil rights activists like the Brooklyn Reverend Milton Galamison in addition to the gripping story of Black Power advocates like Sonny Carson radicalized by white teachers who seemed more interested in expelling "disruptive" black students than teaching them. The public cacophony of so many different voices exhibited an importance beyond the classroom, according to Perlstein:

> The battle over community control, together with the virulent ethnic and racial tensions it unleashed, reshaped New York's liberal politics. It not only exposed growing strains in the political coalition that promoted racial justice and decent working conditions since the New Deal; it gave voice to ideologies that justified liberalism's eclipse. It both reflected and propelled declining liberal hopes that schooling and other government programs could foster equality and promote a harmonious society.[10]

Jonna Perrillo also adds to the story in her book *Uncivil Rights: Teachers, Unions, and Race in the Battle for School Equity*. Her analysis of the events—in which many unionized teachers' lukewarm embrace of, if not outright hostility to, civil rights in New York in the 1950s figures prominently—shows that the labor-liberal coalition may have been even more tenuous than Podair argues it was on the eve of Ocean Hill–Brownsville.[11]

It is important, however, to emphasize two further points. First, the viciousness of the conflict was clearly exacerbated in New York by the developments discussed in chapter 1—the failure of the liberal state to deal appropriately with public-sector labor unions. The two-week teacher strike in 1967—stigmatized with the taint of illegality by an unenforceable law—raised tensions even before the Ocean Hill–Brownsville board's effort to transfer the teachers in 1968.

The UFT's demands in that strike had already pushed them away from community-control advocates. The strike negotiations included a call for an expansion of the UFT's signature school-improvement plan—More Effective Schools (MES), which proposed special schools with low student-teacher ratios and ample remedial services in high-poverty areas—but the UFT also wanted a provision in the contract allowing teachers to remove "disruptive" children from the classroom. As Podair points out, many African Americans, including the Ocean Hill board, viewed a strike under these conditions as a "racial affront—an attempt to withhold educational services to black schoolchildren."[12] The experimental district kept the schools open and attempted to punish striking teachers.

Still, the fact that teachers struck illegally represents an unexamined aspect of the crisis. When the New York Supreme Court upheld Shanker's fifteen-day conviction in October 1967 for violating the Taylor law, it excoriated the strike. Charging the union with "deliberately def[ying] the lawful mandate of the Court," Justice Emilio Nunez asserted that the "strike by a powerful union against the public was a rebellion against the government; if permitted to succeed it would eventually destroy government with resultant chaos and anarchy."[13] An editorial in the *New York Times* connected the union's "wanton resolve to defy the law" with its desertion of "the classrooms that represent New York's first line of attack on the frustrations and despair of the slums." Indeed, the newspaper pointed out the irony that teachers violated the law while striking to demand more power to remove students from the classroom: "The example the union is now giving of self-willed irresponsibility is bound to reinforce the fear already almost universal among parents' organizations, especially those in Negro and Puerto Rican districts, that increased disciplinary power in the hands of teachers would be abused to the detriment of the children."[14]

Indeed, the third UFT strike was by far the longest teacher strike in the city's history; to some observers, it highlighted the utter futility of the Taylor law. Shanker went to prison again for another fifteen days—serving the time in 1969—and New York City's labor mediator (and arguably the most famous in the country), Theodore Kheel, asserted that most of the law should be scrapped. The mediation provisions simply had not worked, Kheel believed, and the ban on strikes was totally unenforceable: "The recent strikes of the teachers represent the complete collapse of the dispute-settling machinery of the law." Further, the labor-relations expert characterized the law's very premises as "schizophrenic" because it attempted to bar strikes while promoting an "ersatz form" of collective bargaining.[15]

The second point to emphasize is that while the strike primarily divided white teachers from black parents and students, it also divided organized labor in the city. Although a large majority of teachers in the city walked out, a significant number of African American teachers (through the African-American Teachers Association and the UFT Black Caucus) opposed the strike (as did some white teachers committed to the principle of community control).[16] Just as it had divided a black community fed up with inferior schools and a mostly white unionized teaching force, the conflict also drove apart different constituencies of organized workers within the labor-liberal coalition.

On the one hand, the national and state AFL-CIO and the New York City Central Labor Council (CLC) officially supported Shanker and the UFT in protecting teachers' due-process rights.[17] Veteran labor and civil rights activists A. Philip Randolph and Bayard Rustin stood by Shanker, too, in spite of the substantial flak they took for the position (Rustin, in particular). Randolph and Rustin knew a good deal about organized labor's incomplete support for racial equality: Randolph in 1960 cofounded the Negro American Labor Council (NALC)—a group integral in organizing the March on Washington in 1963—in response to the AFL-CIO's indifference to black workers.[18] Randolph and Rustin argued, however, that due process had to be protected at all costs. In an advertisement sponsored by the A. Philip Randolph Institute (Rustin was the executive director) in New York City's African American newspaper the *New York Amsterdam News*, Rustin argued that African Americans' access to due process represented a significant victory for the civil rights movement, and any community-control experiment should not destroy those protections: "It is the right of every worker to be judged on his merits—not his color or creed. . . . These are the rights that black workers have struggled and sacrificed to win for generations. They are not abstractions. They are the black workers' safeguards against being the 'last hired and the first fired.'"[19]

Several public- and private-sector unions with large black and Hispanic memberships disagreed with Rustin's position, arguing that the community control the administration in Ocean Hill–Brownsville demanded and due-process rights for teachers were not incompatible. In October 1968, for instance, Victor Gotbaum, president of the largest AFSCME local in the country—District Council 37—introduced a resolution to the CLC supporting both community control and due process. Leaders of Local 1199 (the New York hospital workers union) and the Retail, Wholesale, and Department Store Union also supported the resolution. That initiative was thwarted by Shanker's ally and CLC president Harry Van Arsdale, but on October 26, one hundred black and Puerto Rican labor leaders officially backed the Ocean Hill–Brownsville board. The dissidents included the president of the organization Randolph had formed in 1960. Cleveland Robinson, NALC president, asserted that "the community will not accept a proposition that the solution is . . . an emasculation of the Ocean Hill–Brownsville decentralization project." By mid-November, fifty black unionists staged a sit-in at Van Arsdale's office to bring an end to the strike.[20] Indeed, the labor-liberal coalition did not just fray along the lines of teacher and community during the strike, but the labor movement also experienced a diminution of solidarity as a result of the New York conflict.

Teachers and Black Power Elsewhere

The strikes in New York drew dramatic attention to the conflict in the Big Apple, but militant teacher unions clashed with Black Power activists elsewhere as well. In Philadelphia, a series of smaller skirmishes between unionized teachers and African American activists pushed the PFT away from much of the black community. Between 1940 and 1970, Philadelphia's black population increased by 160 percent while the white population declined and the population of the overwhelmingly white suburbs increased by more than 125 percent. By the mid-1960s, black students comprised 53 percent of the city's highly segregated school population; over half of Philly schools were made up of 90 percent or more of either whites or blacks. By the mid-1960s, Philadelphia's black children suffered disproportionately from the "extensive overcrowding" of the schools and "the highest dropout rate of the nation's ten largest cities" as well as "a large number of obsolete school buildings in deplorable condition."[21]

As Matthew Countryman shows in his study of Black Power in Philly, the Black People's Unity Movement (BPUM) helped organize students in the city for more black teachers and a curriculum more useful for black students. In the most spectacular of these demonstrations—in November 1967—hundreds of black students marched from a vocational high school in South Philly to the board of education's center-city headquarters. Philadelphia's notorious "law-and-order" police chief Frank Rizzo led a fierce repression of the demonstration. The school superintendent, however, sought conciliation by meeting with black students. PFT president Frank Sullivan responded in terms much closer to Rizzo, arguing in an open letter that the superintendent failed to ensure discipline in the schools.[22] In late 1967 and 1968, the union argued that the Pennsylvania Human Rights Commission's efforts to integrate Philly's public schools, in part through teacher transfer, violated the union contracts. In late 1969, the PFT also successfully defended a white teacher—George Fishman—whom students accused of racism at West Philadelphia High School. Finally, black students publicly wondered if teachers, not punished after striking illegally in 1970, benefited from a "double standard" because they were white. When students at West Philly had demonstrated against Fishman, they were ordered to return to class by the courts. In a story reported in the *Philadelphia Tribune*, West Philly High's student-body president asked "if there are one set of rules for the students and another for the teachers."[23]

Teachers and Black Power activists also clashed over decentralization in Detroit. From 1960 to 1980, 25 percent of the Motor City's population left. Middle-class whites made up a large majority of the exodus. The black population increased from 29 percent of the city in 1960 to 44 percent in 1970—a trend that would continue in the years after—and, as Jeffrey Mirel argues, "geographic and social isolation of blacks in Detroit worsened so much that, in 1980, the Detroit metropolitan region was one of the most racially and economically segregated cities in the country." The proportion of black students in the schools increased dramatically while property values declined, making it difficult to maintain funding.[24]

In this context, Black Power activists pushed for community control of schools, arguing that black schools needed black teachers, and collided with the Detroit teacher union in 1967. The DFT president—Mary Riordan—stuck to the position that "color is beside the point if the person is doing the job."[25] When teachers went on strike in 1967 (see chapter 1), they did so just a month after Detroit had emerged from an urban insurgency that left dozens dead and caused millions of dollars of property damage. Striking teachers faced major opposition from the NAACP and other black civic groups, who in these conditions craved nothing but a return to normalcy in the city. Finally, while the state legislature eventually passed a decentralization plan for the schools—which broke the school board into a central board that controlled finances and eight regional boards with some local autonomy—critics skewered the watered-down version they argued was too accommodating to the DFT by ensuring that their collectively bargained contract would not be disrupted.[26]

The Chicago Teachers Union (CTU) clashed with Black Power advocates during this era, too. The CTU—formed when several Chicago AFT affiliates merged as Local 1 in 1937—had, at its inception, the largest union membership in the country at 8,500. Although the UFT would pioneer an official bargaining relationship, the CTU was incorporated into the Chicago Democratic machine and enjoyed "virtual" bargaining rights by the 1940s. The CTU embraced the demand of equal pay for all teachers (the city's overwhelmingly female elementary school teachers had been paid less than high school teachers). Like the New York local, the CTU supported both anti-Communism and color-blind antiracism in the 1950s, and the board of education recognized the union as the sole bargaining agent for city teachers in 1966.[27]

Chicago followed a similar demographic trajectory as Philly, New York, and Detroit. In 1940, blacks represented 8.1 percent of Chicago's population (about 277,000); twenty years later, there were over 800,000 black residents, or 22.8 percent of the city's population. The African American school-age population more than tripled, from 74,000 in 1950 to a quarter million thirteen years later; 84 percent of black students attended schools that were at least 90 percent black, and 86 percent of white students attended all-white schools. Furthermore, black schools were extremely overcrowded when compared to white schools, and the board of education enforced a "neighborhood schools" policy that enforced the color line in the city.[28]

The CTU was lukewarm to civil rights activism, failing to support black boycotts of the schools in 1963 and 1964; the union's president—John Fewkes—repeatedly denied the school board's role in enforcing segregation, emphasized the problem of "disruptive" behavior by black students, and defended teachers' seniority rights against integration of the city's teaching force. Black Power activists took aim at the deplorable state of education for African Americans in the Windy City in the late 1960s. Although the union continued to disregard some of the demands of activists, historian John Lyons argues that the effect on the union's strategy was more positive than in New York City. There were two reasons for this development: first, by the end of the 1960s, about 30 percent of the teachers in the city were African American (more than in New York, where the teaching force was about 90 percent white), and, second, the major voice for African American teachers in the union—the Black Teachers Caucus (BTC)—did not leave the CTU (as the African-American Teachers Association had left the UFT in New York). The BTC instead worked from inside the union to push leadership to seriously consider black concerns.[29]

The two-day 1969 strike featured conflict between black militants who wanted the union to focus on education improvements and white teachers who more highly valued pay raises. A disproportionate percentage of the CTU's black teachers crossed the picket line (45 percent, compared to the 23 percent of total teachers who refused to strike), and this forced the CTU to engage calls for better schools. In the next contract—following a short strike in the same month Pittsburgh teachers walked out in 1971—the CTU won not only higher pay, but also emphasized improvements in education at the negotiating table: additional teachers' aides and smaller class sizes. Almost all black teachers supported that strike, and by 1972 the percentage of black teachers in the city increased to 35 percent.[30] Nevertheless, although the trajectory of Black Power in Chicago ultimately led to more positive developments than it had in New York or Philadelphia, tensions between Black Power advocates and the CTU did not go away.

Newark: A Community Divided

Though conflict existed in many of the cities at the center of the labor-liberal coalition, New York and Newark represented the two most drastic. As we have seen, there is a wealth of excellent scholarship on Ocean Hill–Brownsville. Like the crisis in New York, two strikes in Newark, in 1970 and 1971, exploded in a conflagration of both symbolic and real violence. Unlike Ocean Hill–Brownsville, however, the events in Newark have been sparsely documented.[31] Indeed, in the early 1970s, no single series of labor conflicts went so far as to highlight the divisive terrain of teacher strikes as those in Newark. Revolving around a stark racial conflict, street violence, and mass arrests of teachers, the dramatic struggle attracted serious national media attention, and the AFT tapped even more by publicizing the New Jersey jail term for its president David Selden that resulted from the 1970 teacher walkout.

Teachers had been organizing in Newark since the 1930s, and the AFT granted a charter to Local 481 in 1936. In the 1940s, the Newark Teachers Union (NTU)—like its counterpart in Gotham—was a left-led union with a sizable Jewish membership. Like the CTU, the Newark local pushed the school board in the 1940s to end the pay differential between high school and elementary school teachers. Inspired by the successful organizing of the UFT, NTU activists pushed for collective bargaining in the early 1960s. The NTU particularly attracted Italian American teachers, many of whom came from union families, during this period. According to oral histories collected by Steve Golin, teachers' unpaid "professional" duties—over which they had little say—loomed large in many teachers' decisions to join the union. Also, in the 1950s and '60s, the school population in Newark increased dramatically; teachers remembered class sizes of up to forty during these years and recalled not being able to teach effectively as a consequence. Union organization during these years intensified—in the 1950s, NTU increased its membership annually by an average of fifty-eight teachers; from 1962 to 1966, that number increased to 290 a year.[32]

In 1965, the Newark Board of Education agreed to a collective-bargaining election between the NTU and its NEA rival Newark Teachers Association (NTA). The NTA won the initial election and negotiated the first teachers' contract in the city. The NEA affiliate—more conciliatory than the NTU and "never able to develop a coherent strategy for wresting significant power from the Board"—fell from favor by 1969. Newark teachers worried about violent incidents in classrooms and teacher salaries that were the lowest in Essex County. The NTU also appealed to some teachers interested in making the school system more responsive to the city's poor (overwhelmingly African American) students by pushing the AFT's MES proposal. In 1969, 2,020 of the district's 3,400 eligible teacher voters chose the NTU to represent them at the bargaining table. Only 571 voted for the NTA.[33]

In 1970, the NTU used the election as a mandate to aggressively push for higher salaries and more education resources for the schools. In negotiations, the board offered a $1,300 across-the-board increase to teachers, while the teachers sought a package in which starting salaries rose by $2,000 more than that in order to make the teachers the best compensated in the county. The union also prioritized smaller classes and the implementation of MES schools. Finally, the NTU wanted a formal grievance procedure that included binding arbitration and freedom from workplace duties outside the classroom.[34]

On February 1, teachers voted to strike, and the next day, almost 3,000 of Newark's 3,800 teachers failed to report to classrooms, keeping more than 70 percent of the city's 78,000 students out of school. Like the Taylor law, New Jersey's 1968 revision of its labor statute—the Public Employer-Employee Relations Act—gave public-sector workers the right to unionize and collectively bargain and set up a mediation process. It also outlawed strikes. The school board obtained an injunction, but teachers went on the picket line anyway. Three days into the strike, Theodore Kheel came to Newark to help end the impasse, and a local Essex County judge ordered the arrests of three strike leaders—including NTU president Carole Graves—for contempt. The latter move represented a departure from teacher strike negotiations elsewhere: authorities sometimes imprisoned union leaders—like Shanker, after both 1967 and 1968—but they rarely jailed them during negotiations. Two days later, the judge ordered the arrest of four more union leaders.[35]

As in New York, the strike was intimately related to the racial politics of the city. Since World War II, the African American population had increased dramatically in the city, while the white population declined significantly. About 40 percent of the city's white residents—including many of the teachers—moved to the suburbs. In the late 1950s and '60s, Newark lost a lot of manufacturing jobs—about 25 percent.[36] Further, Newark—like Detroit—had been the site of a violent inner-city insurrection in 1967. Responding to the beating of a taxi driver arrested for driving on a suspended license, black Newarkers—fed up with police brutality and urban poverty—looted and damaged property. The mostly white state police and National Guard savagely put down the rebellion; twenty-four blacks and two whites died, 1,100 people were injured, and police arrested 1,400.[37] The uprising represented a turning point, as Black Power activists, led by Newark's most famous resident, poet and playwright Imamu Amiri Baraka—formerly LeRoi Jones and often still referred to as such by the media—waged a bruising battle to wrest political power from a white city leadership.[38]

Although NTU president Graves was an African American woman, most of the union's leadership and a majority of the rank and file were white. Viewing the strike as against the community interest, Baraka and CFUN organized parents, students, and black teachers opposed to the union to keep the schools open. Carefully steering clear of explicitly characterizing the conflict in 1970 as an outright racial one, Baraka argued that the conflict centered on the "suburban" attitudes of the teachers, many of whom did live outside the city or at least outside the neighborhoods where they taught, and their lack of responsiveness to the "urban" problems of black Newark schoolchildren.[39]

On February 14, city police arrested forty-three rank-and-file demonstrators outside city hall. Four days later, sheriff's officers arrested another thirty-six teacher demonstrators, including Selden, who had come to Newark to support the strikers. By February 22, almost 200 teachers had been arrested, and one report asserted that "exchanges between parents and teachers almost reached the point of hysteria" as the conflict "resurrect[ed] the specter of 'another 1967.'"[40] The board largely gave in: teachers accepted the district's salary offer, increasing starting pay from $6,700 to $8,000 annually, and the board agreed to binding arbitration, class limits of thirty students, and the creation of one MES school.[41]

The new contract expired on January 31, 1971, however, and the same issues from the 1970 strike carried over into the next round of bargaining. This time, however, the political landscape had changed substantially. By then, Baraka and his allies had helped elect Kenneth Gibson as Newark's first black mayor in November 1970. No black nationalist, Gibson nonetheless understood that his victory had relied on Baraka and CFUN, and so he appointed Jesse Jacob, a staunch Baraka backer, as the president of the Newark Board of Education. Jacob had actively worked to defeat the NTU in 1970 and described the contract that emerged from it as "outlandish." He viewed the contract talks as an opportunity to retake control of the schools from the union.[42] Jacob's board demanded that salaries remain at their current levels and that teachers concede binding arbitration in addition to reassuming supervision of children outside the classroom.

The contention over the latter stipulation perhaps more than anything else shows the enormous chasm between the goals of teacher unionism and Black Power in American cities in the late 1960s, and a vital component of this conflict in Newark involved assumptions regarding gender roles. The teachers saw supervision outside of the classroom as an assault on their professionalism; indeed, principals had for years before collective bargaining expected dominion over the teacher's workday and meted out non-instructional chores as punishment. However, as Charles Hamilton pointed out in the piece discussed at the beginning of this chapter, Black Power envisioned the school as central to a broad reconfiguration of community life. Baraka would excoriate the teachers' assumptions about professionalism, writing in April 1971 that "the union says that its members should not have to perform so-called 'nonprofessional' chores, such as bringing children from classroom to cafeteria, or from school bus to classroom. . . . If the union does not even want that human a relationship with the children, why should it want to teach them in the first place?"[43] The gendered nature of Baraka's stance here seems apparent, particularly since much of what he called for represented additional (unpaid) care labor in the overwhelmingly female elementary school domain.

In addition to maintaining arbitration and exemption from nonteaching duties, the union demanded $10 million to increase teacher salaries. Fearing another strike just one year after the last, union leaders waited until February 1 for a strike vote. The vote authorized a new walkout, and because the injunction from 1970 was still in effect, the board again pursued legal action. Nonetheless, as late as February 4, the *Newark Star-Ledger* reported that the two sides had reached a "tentative agreement," and both Graves and Jacob expected a quick end to the strike.[44] Mere days after such optimism, however, they became deadlocked and, once again, authorities arrested NTU leaders. As it had the previous year, the Newark strike gained heightened national prominence in mid-February when AFT president Selden (who had served a forty-two-day jail sentence for violating the injunction in 1970) and vice president Shanker (who stirred up particular controversy because of his connection to Ocean Hill–Brownsville) came to town for support.[45]

On February 24, superior court judge Samuel Allcorn imposed a fine of $50,000 plus $7,500 for each day that the NTU remained on strike. The next day, calling their actions "strikes at the root of the democratic system," he sentenced Graves and two other union executives to six-month jail terms for refusing to order teachers back to work. Graves outlined her reasons for defying the injunction to NTU teachers by appealing to visions of both racial and gender equality: "I stand before you, a black woman, strongly committed to a cause which will bring about unity among black and white workers . . . a cause that will lift the yoke of oppression from around the necks of the poor, both black and white."[46] She eventually served about a month and a half in jail, spending much of the last few weeks on a hunger strike.

For Baraka, the strike represented the furthest thing from uniting white and black workers. Indeed, the activist poet clearly understood the 1971 strike as a racial one between "the Black community and the white suburban teachers' union, over who will control the school system." At the beginning of the 1970 strike, he could at least glimpse the possibility of union and community working together to improve the schools. By 1971, however,

Baraka believed that "the majority of NewArk's teachers share neither space nor values with the Black community of NewArk, plus the fact that they are 75 percent white."[47] Even Bayard Rustin, who—as he had in Brooklyn in 1968—spoke in support of the teachers during the strike's second month, put the stakes in stark racial terms, lambasting the black members of the school board for "acting out the tragedy of former slaves acting like slave masters" in trying to break the strike.[48]

City leadership also split along racial lines. Board president Jacob publicly feuded with white city council president Louis Turco, an Italian American who supported the strikers. Turco had opposed the 1970 strike as too expensive, but his calculus changed by 1971 as he—like the rest of Newark—understood the strike as a conflict over Black Power and the future of the city. At a board of education meeting on March 1, Jacob accused Turco of "talking out of both sides of his mouth." In a speech to PTA groups the next night, Turco called for Jacob, whom he believed was "more interested in a fantastic struggle for power" than schoolchildren, to resign as board president.[49]

The animus spilled over into physical violence. Almost half of the city's teaching force crossed the picket lines, as many of the city's black teachers allied with CFUN instead of the NTU. Black militants from the community worked with Baraka to forcibly keep picketers from schools and assaulted striking teachers, while union teachers physically attacked those on their list of scabs. The union also sought its own muscle, allying with the forces of Tony Imperiale, an Italian American vigilante—and Baraka's arch-nemesis—who, in the wake of the mayoral election of 1970, "announced he was forming a conservative organization to fight Gibson and seeking help from supporters of George Wallace."[50]

Television news highlighted the racial divisiveness and violence brought on by the crisis. On February 11, for example, *CBS Evening News* anchor Walter Cronkite called the strike "bitter" before introducing Robert Schakne's report from Newark. During the five-minute feature, Schakne highlighted the intense racial rhetoric, and images showed Turco explicitly blaming "black nationalists" for trying to break the union and reporting on a "firebombing" of the NTU headquarters. In a barely veiled editorial comment critical of Gibson, Schakne pointed out that "some say he listens too much to LeRoi Jones."[51] A report later in the strike by Ron Milligan for *ABC Evening News* showed images of black schoolchildren on whom teachers had "imposed" a "vacation." The report then cut to a white striker with a bandaged head while Milligan's voice-over charged that "thugs beat up a group of teachers outside their headquarters last night."[52]

On March 25, 200 black Newark school students staged a sit-in at the motel where representatives from the union and the board met, and the strike reached its thirty-sixth day, becoming the longest teacher strike in a major U.S. city.[53] By April 6, board members believed they had reached an agreement, as it appeared that five of nine board members would accede to a mediator's decision already accepted by the union. Ominously, however, the school board could not even vote on the proposal because fights broke out among the 1,000 people crammed into the city hall to watch the proceedings. One reporter characterized the meeting as "Shea Stadium just before the first inning, with the crowds shouting, screaming, and saluting with fists in the air." Police escorted the board members to their seats, and both whites and blacks shouted "racial slurs" at each other. One black commentator asserted that board members should vote against the contract because "we would not like the . . . teachers to return to school." The meeting adjourned when fistfights broke out between white and black Newarkers.[54]

When a vote did take place the next night, the board of education—exclusively along racial lines—voted the contract down five to four. Jacob was the tiebreaker, and he recalled Martin Luther King's March on Washington speech of 1963 to justify his vote, proclaiming, "If this is to be the year of attrition, let it be. . . . Free at last. Free at last. I vote no."[55] The strike dragged on for another week; recalling the bad blood from the insurrection of 1967, Baraka publicly accused the police commissioner of brutality against black activists, and 500 white parents and their children marched on the board of education offices to urge them to stop listening to Baraka and CFUN.[56]

Events continued in this manner for another week, and on April 16, Ron Milligan's report for *ABC Evening News* pointed out that the strike had "virtually torn the city in half."[57] Finally, both sides agreed to Mayor Gibson's "appeal to reason" compromise proposal on April 18, ending the strike of almost three months. In return for agreeing to some nonprofessional activities, the union maintained binding arbitration. Neither side asserted that they had won, and each reacted with relief, not elation. Graves, just out of prison, best summed up the conflict's meaning when she told a group of teachers that "surely no other teachers' strike has been as ugly and as bitter as what we have gone through in the last two years."[58]

Threats to the American Family

Aside from publicizing the physical violence and street fights in Newark, national media focused on the issues of incarceration and law and order, and how the tension between public-sector labor rights and teacher power upset American families and gender roles. Indeed, the 1970s were a time in which fear of national decline and cultural listlessness were tied to threats to the American family.[59] A feature in the January 1971 issue of *Redbook* showed that teacher strikes represented another arena in which turmoil threatened conventional gender roles. The piece featured Betty Rufalo, a white mother of two who had been one of the 200 teachers arrested in the 1970 strike. Rather than appeal her sentence as had most of the teachers, Rufalo had chosen to serve a thirty-two-day sentence in the Essex County Penitentiary to make a political statement. The author of the article, Dorothy Gallagher, argued that Rufalo's imprisonment showed that America had difficulty discerning the legitimate citizen from the criminal:

> There was a time when we knew without thinking about it who our criminals were. They were murderers, rapists, thieves, and we were certain that their acts, whether or not they were rooted in the soil of our society, could not be tolerated. It is not easy to have that kind of certainty about many people who are called criminals today: the draft resister who refuses to fight in what he considers an unjust cause, the civil rights advocate who challenges a segregationist law. . . . Have they too committed crimes against the people?

Although the rhetorical question espouses sympathy with Rufalo's choice to go to jail, Gallagher's piece also described how Rufalo's militancy and violation of the law challenged notions of female propriety, including her duty—as a union activist willing to go to jail—to her own family. After defining Rufalo as a "wife, mother, elementary school teacher," the article described her prison ordeal, which began with her stripping off "all of my clothing except my panties and bra." The feature then characterized Rufalo as "fighting . . . to have some tenderness" for her children on the outside, to maintain her role as a mother: "'The other girls,' [Rufalo] says, 'got letters that their children were in the hospital and they didn't shed a tear; they're in worse hell than everyone else and knew they had to take care of themselves first.'" The illegal actions like those in Newark that put women in prison represented one more place in which Americans questioned the gender roles that had been the foundation of the ideal family.[60]

In December 1971, Newark was in the national news again. All of those NTU teachers who—unlike Betty Rufalo—had appealed their jail sentences for violating the 1970 injunction were finally due to serve them. Most had chosen, as a symbolic gesture, to serve their sentences over Christmas vacation rather than deprive their students of education days, and NBC's Jim Collis was on hand to report on the spectacle of 128 schoolteachers beginning nine-day prison terms. The reportage was sympathetic to the teachers, but like the *Redbook* article on Rufalo, it suggested that the teachers' militancy had disrupted the American family. Collis's voice, layered over images of hugging family members, stressed that "this was a mixture of bright, earnest young men and women and gentle-looking people old enough to be grandparents." An image showed a white father holding a baby, offering that he felt "kind of crummy" about not seeing his child for the holiday. A black mother indicated that "I have a child who's going to be very lonely without me this Christmas." Collis's follow-up story on December 31, as the teachers emerged from jail, showed images of reuniting families and one of those "gentle-looking grandparents" remarking on how "wonderful" the fresh air was, and pointed out that many of the teachers "still face[d] sentences from the 1971 strike."[61]

This last statement, in addition to underscoring the ways that teacher militancy disrupted the family, also showed that, in its casual reference to the 1971 Newark strike, the reporter assumed the viewers' familiarity with the conflict. Indeed, it seems clear that by the end of 1971—as a result of Newark as much as Ocean Hill–Brownsville—the teacher strike had indeed emerged as a battleground of national political and cultural significance.

Letters to the Essex County Penitentiary

The Newark strikes also crystallized a national political discussion when the AFT publicized its president's sixty-day prison sentence in 1970 for violating an antipicketing injunction. Taking out newspaper advertisements in New York, Chicago, and Los Angeles, the AFT implored readers to write to Selden in jail. When they did, however, Selden found out that the conflicts borne of teacher strikes had upset many Americans' political sympathies.

In American labor history, employers have often sought injunctions to break strikes. Invoking property rights, state and local judiciaries almost always sided against workers before the advent of 1930s labor policy. Less frequently, the federal government sometimes issued antistrike injunctions, as when a U.S. circuit court did so against union leader Eugene V. Debs during the 1894 Pullman strike. Although 1930s labor law circumscribed private employers' ability to use injunctions, they still represented a major problem for public-employee unions. Selden was only the second president of a national labor union sentenced to prison for violating an injunction after Debs (the other was Michael Quill of the Transport Workers Union [TWU]). Rather than appeal, the AFT president chose to go to jail—ultimately serving forty-two days—and the AFT used his imprisonment to agitate for public-employee strike rights. The public relations strategy focused on bringing attention to Selden's jail term followed by a series of "bread-and-water" receptions after his release to promote anti-injunction legislation. The AFT bought full-page advertisements in the *New York Times*, *Chicago Tribune*, and *Los Angeles Times*, and the ads concluded by providing his jail-cell number so readers could write to him in prison. The letters rolled in from across the country that April, and, aside from giving Selden interesting reading material while incarcerated, provide a valuable snapshot of the shifting political allegiances in the United States in the early 1970s.

Indeed, Selden's action provoked responses from Americans concerned about many different things, most of which were much larger than his arrest.[62] The prison letters highlighted exactly why debates over teacher labor conflict and public education were so important in the era, as well as how these conversations led some Americans to begin to link urban teachers and working-class African Americans through the racially charged discourse of "law and order." In fact, the letters to Selden indicate an ironic development: dysfunctional metropolitan political economy caused a fundamental conflict between union power and Black Power, but critics of both began to view them as part of an illegitimate, "unproductive" coalition buttressed by a liberal state in victimizing hardworking, law-abiding citizens. For these critics, striking teachers and the urban poor, in tandem, had caused America to lose its way.

When David Selden began serving his prison term in 1970, he had for twenty-five years been at the vanguard of teacher unionization; thus, the AFT president represented a fitting symbol for those who wanted to assess the position of public-sector labor. Selden came of age in Detroit, putting himself through college as an automobile worker just before the famous UAW sit-down strikes in 1936–37. His first teaching job, "in a K–9 school in a slum area in the shadow of the Ford Rouge plant," galvanized him toward activism, and he was elected president of the Dearborn Federation of Teachers just before WWII. Settling in New York City after the war, he organized for the Teachers Guild in the 1950s, and then for the UFT; indeed, Selden was instrumental in the 1960 strike. After helping the UFT win its first contract, he worked to organize collective-bargaining efforts elsewhere. Winning the presidency as the AFT's Progressive Caucus nominee in 1968, Selden's platform focused on more school appropriations, smaller class sizes, and better teacher salaries. He had almost completed the first of three two-year terms when he began his jail sentence.[63]

In trying to change the minds of the public, the AFT attempted to counteract the growing antiunion stance of many newspapers that colored coverage of teacher strikes. Labor's growing strength in the 1930s led to the development of a "labor beat" by many newspapers that shaped public opinion toward unions in a sympathetic direction. By the end of the 1950s, however, labor reportage had increasingly included sensational exposés of supposed union excess and corruption by journalists such as Westbrook Pegler and Victor Riesel.[64]

Teacher strikes—since they kept kids out of school—lent themselves well to such exposés. One long-form example can be found in the 1972 book written by Robert Braun—a journalist for the *Newark Star-Ledger*—titled *Teachers and Power: The Story of the American Federation of Teachers*. The beat reporter behind much of the coverage for New Jersey's largest newspaper on the Newark strikes, Braun's reportage included analytical pieces showing that the students were the real "losers" in the strike. Still, while he criticized teachers, he did not argue that they were solely at fault. Although the NTU, in his opinion, viewed the "traditional trade union contract . . . with more sanctity than the right of a child to come out of poverty through education," he equally criticized the board of education for "frightening teachers" into striking.[65]

By the time the book came out, however, Braun presented the 1970 strike as the central evidence of a damning critique of the excessive power he believed unionized teachers wielded. Braun's book did not yet evidence the antigovernment animus that would place teacher unions within the context of a more general attack on the public school system by free-market conservatives later in the decade.[66] Instead, Braun contended that union leaders cared only about accumulating power, thus abrogating their duties as teachers, or "transmitters of knowledge and all that is best in American culture."[67] In summarizing the growing phenomenon of teacher strikes in the late 1960s and early '70s, then, Braun's thesis advanced the idea that unionized teachers were destroying the education system. "The union operates the schools without any genuine responsibility to anyone, certainly not the public," Braun argued, concluding that "[it] may very well take away the last chance that the people of this nation have for a truly free and public school system."[68]

Labor conflicts in the newspaper industry had also eroded support for unions in both reportage and on editorial pages. The Newspaper Guild strike of 1962–63 in New York City, which virtually blacked out print news in the city for several months, caused multiple newspapers to permanently cease production and raised tempers among the remaining publishers in the Big Apple. By the early 1970s, newspapers across the United States faced profit crunches and struggled to compete with television news and, in urban markets, to maintain readerships after suburbanization shrunk the pool of public-transit commuters. Collectively bargained contracts represented an impediment to cutting costs, and efforts by newspaper unions to fight mechanization caused some publishers to criticize all unions.[69] The advertisements, then, represented an interposition of the AFT's view into an increasingly unwelcoming environment.

Unsurprisingly, Selden received numerous letters of support from teachers "inspired" to organize and even from unionized educators in Latin America and Europe.[70] What really stands out in the archive, however, is how many people wrote who were not teacher unionists. Of the letters he received in March and April 1970, thirteen letters of support came from those who were not teachers or union activists in addition to seventy letters criticizing him. Of the latter, twenty-one were newspaper ads defaced in some way and then sent to Selden.

Many of those who supported Selden but who were not teachers focused on the injustice of his imprisonment. R. E. Moore, from Richmond, Virginia, felt "disgusted and heartsick over something which I had no idea could happen in my own country." Moore's twelve-year-old daughter aspired to teach, but after he read about Selden, he wanted to "make [her] sit down and read this article after which her mother and I will begin trying in any way we can to correct this situation . . . or, I guarantee you, my daughter will not be allowed to persue [sic] a career in teaching."[71]

Selden's detractors believed just as fervently that the union president was wrong, and many expressed these beliefs through vicious racialized and gendered ad hominem attacks. James Arbuckle, from Massachusetts, made out a check for "$0,000.00" to Selden, whom he called a "Christmas turkey." Identifying himself as a member of "Agnew's Silent Majority," another critic named Selden a "parasite," explicitly comparing him, in racially charged phrasing, to the "professional poor and the welfare cadillac group." Al Eischen, from the Chicago suburb of Des Plaines, Illinois, called Selden "self-righteous," while an unsigned letter—highlighting the gendered assumptions about a predominantly female union with male leaders—asked that Selden "please try to act like a man, even if you can't be one."[72]

Rick Perlstein's study of the Richard Nixon administration has shown that period to include "the rise . . . of a nation that had believed itself to be at consensus instead becoming one of incommensurate visions of apocalypse: two loosely defined congeries of Americans, each convinced that should the other triumph, everything decent and true and worth preserving would *end*." In Perlstein's formulation, Nixon not only reflected this larger irreconcilability, but also served as its provocateur par excellence.[73] The letters to Selden, however, show that much of the vitriol so prominent in the political discourse in the late 1960s and '70s did not come from the top down. National figures clearly shaped the rhetoric of the era, and references to the "silent majority" in the letters prove that ideas flowed from political elites to ordinary Americans. The fact that so many individuals from different parts of the country, however, responded so vehemently to Selden shows that this discourse also welled up from the grass roots. Further, as the juxtaposition of the "silent majority" and the "parasites" indicated, these "congeries" imagined by Perlstein were beginning to break down along the lines of the "productive" and the "unproductive."

Unions, Professionals, and Victims

Other Americans wrote to Selden because they believed strikes had highlighted the excessive reach of public-sector unions who victimized hardworking Americans in the process. When David Perry saw the AFT's advertisement in the *New York Times*, for example, he drafted a letter that very day. He typed it on letterhead from a company called Automatic Information Dispatching Systems based in West Orange, New Jersey, a suburb of New York. Perry argued that the Newark teachers, in violating an injunction, had acted in a more "'union' oriented approach" than an "'educator' oriented approach." The implication seems evident: it might be okay for blue-collar workers to form unions and strike, but for "professionals," it was undignified.[74] Because of developments at both the national and local level, then, individuals like Perry viewed with skepticism the notion that professionals should use the same tactics as working-class people.

Those who wrote Selden also showed that his example caused them to believe that public-sector unions had corrupted government. Ralph Curcio, a New Yorker, drafted a letter on April 15. He began by sarcastically exclaiming, "Too bad!" to Selden's imprisonment. "Injunctions," he lectured the AFT president, "serve a useful purpose in our society—to prevent strikes from dragging out ad infinitum." Connecting them to corruption elsewhere in municipal and state government, Curcio asserted that "teachers as well as other public officials have taken unfair advantage of people in the past by stubbornly demanding the last penny." He concluded by pointing out that "fair compensation for anybody's work, whether teacher, policemen, sewer cleaner, laborer or whatever is certainly reasonable. But piggy-ness and greed have no place in contracts."[75] The letter's fundamental point—that adequate compensation arises naturally and that public-sector workers extort the public—gives us a clear window into one of the basic contradictions around public-sector labor bargaining in the 1970s. In spite of Curcio's characterization of them, teachers had only made significant wage increases through militant action because it had taken either strikes or at least the threat of them to receive the "fair compensation" to which he believed all workers were entitled.

John Amber, from Marengo, Illinois, also focused on the victimization of Americans by public-sector unions and implied that the state had abrogated its duty to protect those aggrieved by union excesses. Amber argued that "there are other means of achieving desired ends rather than reducing municipalities, states, and the nation to helplessness brought about by striking public workers."[76] For this reason, he applauded Selden's sentence, as the state finally appeared to be protecting *his* interest, even if it seemed a little too late. In the 1970s, conservative think tanks like the Hoover Institution and the American Enterprise Institute increasingly worked to further the argument that both corporations and the white working class suffered similar oppression as victims of the liberal state.[77] Amber's response to Selden's imprisonment certainly indicates the plausibility of such rhetoric; he obviously believed that a state largely unconcerned with such victimization abetted the predations of public-sector unions.

Teacher Strikes and Law and Order

Barbara Mancbach, writing on April 14, found the AFT's public relations campaign "truly sad." She lectured Selden that he was in prison because he "willfully disobeyed the law," and she believed the stakes of his actions were higher than the injustice of the injunction: "I too, feel that many laws discriminate against me, both economically and personally. Millions of other Americans feel other laws are unfair to them. If we all disobey these laws we disagree with we will have anarchy." As her letter continued, Mancbach pointed out to Selden that he and the striking teachers set a poor example for students: "You've taught them how to disobey the law. Pick up any paper and read what's happening on college campuses throughout the country. The students have learned well the teachings of their teachers." Indeed, for Mrs. Mancbach, the teachers' actions contributed to her growing anxiety that the United States had developed a generation of people without respect for the established legal and moral structure of the nation.[78]

Mancbach was not alone. Like her, many others presented a view of Selden not as an overreaching labor leader but as an immoral teacher. Just as some wrote to Selden as a vessel for their shifting views of labor unions, others saw him—in his role as a teacher—as an arbiter of cultural values. Going back to the nineteenth century, teachers had been central to the mission of American public education, an endeavor in which, in historian William

Reese's words, schools were expected to "strengthen the moral character of children, reinvigorate the work ethic, spread civic and republican values, and along the way teach a common curriculum to ensure a literate and unified public."[79] If the role of schools had changed dramatically from the onset of the common school movement, the notion that teachers were integral in inculcating American values had not. If, as Barbara Mancbach believed, respect for the law defined legitimate American culture, Selden symbolized its failure to be transmitted to the next generation.

In response to student protests and inner-city uprisings by African Americans against police brutality, political figures as diverse as Richard Nixon, George Wallace, and Frank Rizzo stoked fears of a breakdown in law and order to political advantage. Rizzo—the Philly police chief—made his name through violent suppression of Black Power protests; he eventually won the Philadelphia mayoral election in 1971, serving two terms.[80] Wallace's ability to profit politically from highly racially sensitive issues like school busing and urban violence have been well documented.[81] Co-opting some of Wallace's popularity, Nixon used a milder form of "law and order" as a key component of his bid for the presidency, and implicit in his famous reference to the "Great Silent Majority" in his November 1969 speech was that support for the Vietnam War lay with the bulk of those law-abiding citizens who disapproved of the student movement's protest tactics.[82]

Illegal teacher strikes like those in Ocean Hill–Brownsville and Newark took place in this context. As we have seen, both UFT president Shanker and his NTU counterpart Carole Graves served jail time for leading strikes. For many Americans frustrated by the sense that law and order had degenerated, a teacher willing to defy an injunction seemed to be more a criminal—similar to the insurgents in cities like Newark where, ironically, Black Power clashed with teacher unions—than a martyr. Indeed, many who wrote Selden disapproving of the teachers' tactics defined themselves, in contrast, as law-abiding citizens. John McKay, who disfigured a *New York Times* advertisement from April 19, 1970, for example, asserted that "Selden broke the law, & accordingly, should have to pay the penalty—if <u>we</u> broke the law, we would have to pay."[83] In another defaced advertisement, this one unsigned, the writer called herself/himself a "quiet, law-abiding citizen." She/he pointed out that "government employees are stopped by law from striking. They may resign get another job but not strike against me and you and John Doe."[84]

Other commentators used racially charged rhetoric to make the point. In coded reference to the wave of urban uprisings in the late 1960s, a defaced advertisement argued that "RIOTS should not be [teachers'] 'thing.'" Edward Withing from Caldwell, New Jersey, believed that "it is no wonder our young people riot, burn and destroy with your group showing the way to disobey law and order." Jack Sherman, from New York City, agreed. He wrote Selden that "one of the reasons for all the violence and disorder that is going on in this world is because our teachers, who are supposed to teach our children to respect and obey the law, instead . . . show our children how to become criminals."[85]

Reappropriating an advertisement in the *Chicago Tribune*, another commentator celebrated the AFT president's imprisonment. Condemning Selden, student radicals, and urban African Americans, she/he wrote next to Selden's picture: "CONGRATULATIONS: HOPE YOU SPEND ANOTHER 36 DAYS THERE. WE'RE SICK OF PROTEST DISSENTERS, RIOTS & ALL OTHER LEFT WING ACTIVITIES."[86] Yet another defaced advertisement rewrote teachers into a larger narrative of illegal union corruption. This critic suggested how Selden might make full use of his sentence: "Congratulations, Dave. Have fun. Be Happy, Dave. To help pass the time you might initiate a by mail chess game with that other votary of the public weal, Jimmy Hoffa!"[87] At the end of the 1960s—a time of increasing backlash against social-protest movements—Selden's act of civil disobedience represented to some Americans nothing more than the fraud and bribery convictions of the imprisoned Teamsters president.

In making sense of letters from those who believed Selden should have obeyed the law, it is clear that the disorder and unrest purportedly brought about by the tactics of unionized teachers symbolized a larger decline in respect for American moral values. Truly, for many Americans in 1970, the fact that even teachers engaged in civil disobedience had come to symbolize a larger crisis in American society.

Conclusion

This chapter examined conflict over teacher unionization in the late 1960s and '70s that pitted teachers against efforts by Black Power activists to gain varying degrees of political influence over urban school systems. After the failure of liberal labor law to prevent strikes emerged as a critical problem by the late 1960s, teacher conflict became even more deeply contentious as the terrain of this clash centered on African Americans fed up with the gross institutional inequality of many American cities. While these two constituencies engaged in a struggle in which political and economic power appeared to be a zero-sum game, a growing number of critics of the trajectory of labor liberalism had begun employing the heavily freighted concept of law and order to link teachers and the urban poor as the political foil against which they defined themselves. This characterization would only grow in power when some of America's largest cities faced the specter of fiscal crisis.

CHAPTER 3

"WHO IS GOING TO RUN THE SCHOOLS?"

Teacher Strikes and the Urban Crises of 1972–73

In January and February of 1973, teacher strikes closed down Chicago, Philadelphia, and St. Louis. These actions meant that, at one point, schools were closed simultaneously in the second-, fourth-, and eighteenth-largest cities in the United States.[1] Then, when school opened in September, teachers in the nation's fifth-largest city—Detroit—walked out for seven weeks. Almost one million students were idled in the winter of 1973, and another quarter million in Detroit that autumn. Observers in the media groped for a narrative to make sense of the strike wave. On the evening of February 2, for example, *ABC Evening News* featured a story about the walkouts in St. Louis and Philly. Reporter Ron Miller spoke about the phenomenon of teacher strikes: "The growing national militancy among teachers is being addressed to a larger question: power and the traditional roles of education." The report cut to a white teacher in St. Louis (then on strike for almost two weeks) who said teachers were "demanding emancipation . . . we really want to start making decisions about education." The school board president dutifully expressed his skepticism about letting unionized teachers dictate education policy. Finally, Miller concluded the story by predicting that "the snowballing effect of teacher militancy will affect the seat of power in the educational establishment."[2]

The language the reporter used to describe the strikes tied St. Louis, Chicago, and Philadelphia to a larger sense of crisis around the American education system ushered in by the half decade of urban teacher strikes beginning in the late 1960s and exacerbated by the spectacular conflicts in New York City and Newark. Indeed, when teachers talked about "making decisions about education," they meant decisions about working conditions: salaries, class sizes, due-process rights, and even pedagogical techniques. Virtually every aspect of education, however, included budgets and who would pay for them, and these budgets became particularly controversial when cities faced a declining local tax base. For teachers, these questions became especially acute as the economy slowed down, exacerbating deficits, and inflation bit into the gains they had made since the early days of collective bargaining. Further, this larger discussion about metropolitan political economy—following the dynamic efforts of Black Power activists in the late 1960s—had also become inseparable from discussions about racial inequality in addition to the role gender played in determining who made sacrifices for fiscal precariousness. Thus, the three long strikes in particular—in addition to Detroit's lengthy strike, St. Louis teachers struck for four weeks; in Philadelphia, eight weeks—became battlegrounds for raging debates about the fiscal crisis of urban school systems and the trajectory of the American city. By the end, all four of these cities seemed unable either to provide education to its citizenry or to find a workable solution to its structural funding problem.

Indeed, the unique problems associated with deindustrializing American cities and an integrated metropolitan area arbitrarily broken into districts that allowed suburbanites to avoid contributing to the city often figured into the initial discussion of the conflict. As teachers went on strike illegally, however, the public conversations focused more on the intransigence of the teachers and the perceived inability of the government to control the excessive demands of its own labor force. As the strike ground on in Philadelphia, for example, public debates over finding new sources of revenues for the beleaguered city dissipated. In St. Louis, the illegality of the teachers' action also took center stage, and the drastic actions needed to divert some small funds to education highlighted an emergent crisis of the labor-liberal coalition in that city. During the course of the Detroit strike, the discussion increasingly focused—in the context of budgetary limitation and teachers' violation of the law—on ensuring that teachers remained "accountable" for their actions both inside and outside the classroom.

This chapter focuses on these four conflicts—with special attention to Philadelphia as the longest and most dramatic. In a template manifested more dramatically in the New York City fiscal crisis (see chapter 4), the suite of policy options narrowed significantly as the public debate revolved around the behavior of teachers.

Philadelphia: Sounding the Alarm

By the beginning of the 1960s, Americans had become used to an insignificant annual inflation rate of about 1 percent, but by the early 1970s, inflation had become a serious problem. For a variety of reasons—not the least of which was the Johnson administration's spending on Vietnam—inflation increased in the late 1960s, reaching 6 percent by the end of 1969 and more than 6.5 percent by December 1970.³ For teachers, inflation challenged salary gains they made since the onset of the collective-bargaining era and became a major cause of anxiety (and thus, militancy) in the 1970s. Indeed, the average annual salary of U.S. teachers increased from about $5,000 a year in 1959–60 to $17,600 by 1980–81. When adjusted for inflation, however, the average salary increased about 33 percent from 1959–60 to 1969–70 and peaked in the 1972–73 school year. In real value, teacher salaries declined from 1972–73 to 1980–81 by about 15 percent. In fact, the average teacher salary did not return to 1972–73 levels until 1986.⁴

President Nixon, worried about reelection in 1972, also harbored a lot of anxiety about inflation and, in August 1971, instituted wage and price guidelines to slow it down. Labor leaders feared pushing for higher salaries, and, as a result, strikes decreased in the 1971–72 school year.⁵ By the fall of 1972, however, as contracts came up for renegotiation, teachers worried that when the administration lifted wage and price controls, inflation would spike again. The resulting inflation, in fact, would end up rising beyond anyone's worst fears; by the time all of the wage- and price-suppression measures were gone in 1974, inflation, further exacerbated by the OPEC oil embargo in late 1973, jumped back to 6 percent and would reach double digits by September.

In Philadelphia, inflation combined with a budget crisis to lead to a bruising conflict between the PFT and the board of education. The budget shortfall following the 1970 strike had become critical by 1972; the school board, at the opening of the 1970–71 academic year, faced a deficit of almost $50 million and, although teachers had agreed to a slightly longer workday in exchange for a modest wage increase, by the time that contract expired in 1972, the district faced an even wider revenue gap. In a development to be replayed in more dramatic fashion in New York City three years later, Philadelphia banks refused to continue lending the schools money unless the board formulated a plan to balance the budget; without more money, the board estimated that schools would have to close two months early.

The board of education's contract offer in August 1972 seemed certain to provoke a strike. In addition to freezing salaries, the proposal extended the teaching day and eliminated over 500 jobs. Not only did teachers oppose the cuts, but they also asked for enough new positions to reduce class sizes and a 34 percent salary increase for all teachers to offset the 7 percent inflation since the last contract and the expectation of higher inflation when Nixon lifted price controls. Neither side expected to win their proposal entirely, but the enormous gap between the two sides as negotiations began was evident. Making circumstances even more difficult, Frank Rizzo—the former city police chief and a leading proponent of the growing "law-and-order" discourse in the early 1970s—won the 1971 mayoral election on a pledge not to raise taxes.[6] For Rizzo, upholding the law and holding the line against taxes represented the kind of toughness needed to maintain the racial and gender hierarchy in the city. It is not coincidence that Rizzo—himself a public employee in a police force of male blue-collar workers—opted to stand against a union comprised of a majority of women instead of curbing spending on the city's police force.

The board refused the teachers' offer to work under the old contract during negotiations, so 13,000 teachers walked out on September 5, 1972. After three weeks, the board reversed itself and agreed to bargain while teachers worked with a new strike deadline of December 27. For three months, Philadelphians—many of whom organized demonstrations and letter-writing campaigns to the union and the board of education—lived in fear that the schools would be shut down again. By the end of December, the impasse remained. Teachers postponed the deadline another two weeks but left classrooms again on January 8. The school board sought an injunction, arguing that under PA 195, the walkout constituted a danger to the "community welfare." Three days later, a local judge ordered the teachers back to work. The PFT refused to call off the strike, and the union's two top leaders were arrested, convicted of contempt of court, and sentenced to jail terms of six months to four years. They began serving the time immediately.

The strike continued into February, and the PFT, hoping to break the impasse, convinced the Philadelphia AFL-CIO to threaten a one-day general strike: seventy-five labor leaders in the city promised that their members would strike in solidarity. Citing their concern for female teachers, the Philadelphia chapter of the National Organization for Women (NOW) called on its members to observe the demonstration as well.[7] At this point, Nixon sent assistant secretary of labor William Usery to mediate, and Usery got the PFT and the board to agree on a four-year, $93-million contract that would raise teachers' starting salaries from $8,900 to $10,324 by the end of the deal. The new contract averaged a raise of 4 percent a year—increasing the costs of the school district but hiking salaries less than the 5.5 percent maximum pay increase suggested by Nixon's anti-inflation policy and below the actual rate of inflation over the contract's life. Altogether, the two strikes shut down one of the nation's largest school districts for three months. Further, although the State of Pennsylvania committed an extra $24 million to Philly schools and Mayor Rizzo was forced to seek new revenue sources—taxes that would not affect the "working man," in his words—the ordeal failed to resolve the structural issues bedeviling the district in the first place.[8] Most significantly, however, the teachers' violation of the law effaced the citywide discussion about the fundamental flaws of funding schools in Philadelphia that had been present at the strike's outset.

Philadelphia: Defining the Strike

As might be expected, the school board early on attempted to blame the fiscal shortfall on the teachers. At its first meeting after the strike began, one board member argued that teachers needed to be accountable for taking advantage of the school district: "They are going to have to realize the fact that maybe their Santa Claus is dead and that a long strike is not going to resurrect him." Another board member, who pointed out that he voted for the last three "extremely generous" contracts, implored the teachers to sacrifice for the schools. Even William Ross—the board of education president, an executive for the International Ladies Garment Workers Union (ILGWU), head of the Philadelphia Dress Joint Board, and a vice president of the Philadelphia AFL-CIO council—criticized the union's demands: "Two weeks ago, I stated that there was a race on as to who would close the School System down . . . between the union and the banks. I thought that the union would get the hint because the banks said they wouldn't loan us any more money. But the union decided to jump the gun." Ross's stance as a labor leader underscored the potential gulf between public- and private-sector unions over militant teacher tactics: Ross and other labor leaders did not always reflexively support public-sector unions—particularly given the sensitive nature of strikes that could shut down urban education systems.[9]

The local media—both on television and in print—was initially skeptical of the school board's claims. A union town—organized labor was powerful in textiles, construction, and other industries—Philly became a major center of the New Deal coalition with a vibrant working-class culture in the 1930s. Even its public sector had a history of strong organization, particularly its AFSCME local.[10] Most television and newspaper editors who commented on the strike were, if not always union backers, realistic about the reciprocity of labor negotiations. WCAU-TV's (Philadelphia's CBS affiliate) editorial director Peter Duncan, for instance, argued in a piece broadcast during the station's 6:30 p.m. news program that teachers appropriately sought a salary increase because other city workers recently received higher pay. Still, Duncan also argued that this raise "must be within the realm of existing realities." The teachers (and the school board) needed to understand that "any settlement . . . will have to reflect the desperate financial straits in which the School District finds itself." Indeed, Duncan singled out neither side and concluded by asking both to "hammer out" the contested issues at the "negotiating table."[11]

Both of Philadelphia's major newspapers approved of Pennsylvania's new public-employee law and supported teacher salary increases in 1970. Both newspapers' coverage of the 1972 strike, however, definitively connected teacher pay to the city's fiscal future. Indeed, in the early 1970s, each reported extensively on the budget difficulties, and both the *Inquirer* and the *Evening Bulletin* pointed to the school system's deficit and criticized the union's refusal to submit to concessions. With daily circulations of over 500,000 each (and Sunday circulations well above that figure), these two newspapers helped very much to frame the public discussion of the strike.[12] The *Inquirer* placed a good deal more blame on the teachers for city finances, while the *Bulletin*, at least during the September strike, did more to assess the school district's structural limitations.

On September 17, for instance, the *Evening Bulletin*, in response to an encouraging announcement that Rizzo had diverted money from the city budget to the schools and convinced the state government to advance some more, was still pessimistic: "Even if the teacher strike is settled tomorrow (which is highly unlikely) the Philadelphia School District's monumental problems will remain monumental. . . . The money problem is chronic, and the ordeal is very real and desperate." The newspaper editor cited a recent Philadelphia Chamber of Commerce position paper characterizing the strike as "merely compound[ing] an existing crisis." Rather than criticize the teachers' demands, the piece grudgingly agreed with the chamber's call for an increase in the city's property tax and a new business tax. It admitted—an assertion that now seems unimaginable given the right turn by the national chamber in the '70s—that "when Philadelphia's business community sanctions an increase in property taxes and a floor to the business tax, it certainly dramatizes the gravity of the schools' plight."[13]

The *Philadelphia Inquirer*, guardedly optimistic in 1970 about PA 195, was much more critical of the union in September 1972. The *Inquirer*'s editor in chief, Creed Black, penned an editorial at the onset of the strike questioning the wisdom of allowing public-sector employees to strike (Black was assistant secretary for Nixon's Department of Health, Education, and Welfare when the law passed two years earlier).[14] About a week later, an unattributed editorial demanded that teachers agree to concessions in order to "end this senseless and futile strike." The board of education, the *Inquirer* argued, "would be committing financial suicide if it agreed to reopen the schools" without extending the workday and cutting jobs.[15]

On September 30, George Wilson—a different *Inquirer* editor—articulated one of the first openly antiunion critiques of the strike, roundly condemning collective bargaining by teachers. He argued that union leaders wielded too much influence over the board's managerial prerogatives. Too much union power—in both the public and private sectors—was a "major cause of inflation, unemployment, and high taxes." According to Wilson, all unions kept those with the best ability to manage from doing so, and this represented a particularly acute problem in the public sector: "When unions usurp management powers in private industry—by imposing inefficient work rules on railroads, for example—the result is higher prices and a diminished ability to compete with foreign products at home or abroad. When the same thing happens in the public sector—as in the administration of public education—the result is higher taxes to subsidize waste and inefficiency." The major question as the September strike ended, then, was "who is going to run the schools—the Board of Education or the Federation of Teachers?"[16]

Four different public constituencies vied to define the school district's fiscal difficulties in the wake of the September strike. First, teachers argued that conceding hard-won gains was anathema to the logic of collective bargaining. Second, a constituency best defined as liberals who lived mostly in either the white elite, middle-class, or working-class areas of the city viewed education as crucial and worth spending money on. They believed that teachers deserved higher salaries and students deserved smaller class sizes, and though not enthusiastic about the strike, their basic solution to the education crisis was to find new sources of revenue. The third constituency, consisting of white middle- and working-class Philadelphians, had grown increasingly critical of the trajectory of labor liberalism. They allied with the Democrat Rizzo, and lived mostly in the white neighborhoods in Northeast Philadelphia. This group opposed higher taxes and, sometimes, flatly linked this opposition to an aversion for paying for, in their words, "ghetto" schools. The final constituency, living in the predominately African American neighborhoods in West Philly and North Philly, believed in the vital importance of education, but after years of failed efforts to provide decent schools and a teacher union that had often opposed the initiatives of Black Power, viewed the claims of the PFT to improve the education system through higher salaries with cynicism. They attempted to expose what they viewed as the union's racial assumptions and to get teachers back in classrooms.

Philip Rosen, a teacher at Northeast High School, wrote candidly to PFT president Sullivan. He outlined, after an "informal poll" of his colleagues, a reasonable settlement to the September strike. First, he believed teachers should concede the longer school day because "an additional 40 minutes would not be unbearable." Further, "giving in" would be beneficial since it would be "symbolic to many as a form of 'dedication.'" Second, Rosen found—given the "hard times"—a 5.5 percent salary increase "satisfactory." Class size, however, was "most important" and "unnegotiable." Rosen believed the thirty-five-student limit was already excessive and pointed out the potential difficulty of teaching forty or fifty students, as the board proposed.[17]

Rosen's letter underscores two important points regarding the teachers' perspective. First, they did not seek exorbitant pay. The 35 percent figure cited by the board and the media represented part of a dance in which the union opened with an enormous figure and later came down substantially. A union teacher like Rosen understood that the PFT actually sought more modest raises. Second, in pointing to the symbolism of a longer school day, he knew that garnering public support was fundamental in winning the strike.

Following similar logic, other teachers and their family members tried to influence the debate by writing to local newspapers. On September 20, for example, the *Evening Bulletin* published several letters from teachers defending the strike. Barbara Mitchell, for instance, appealed to Philadelphia's masculine labor culture. "Is there a union member in this city," she asked, "who would give up benefits his union strived so hard to get for him?" Sharyn Callahan showed that, as in Newark, more nonteaching duties represented an assault on "professionalism." She argued that if the board cut jobs (some of which were not instructional positions), it would force her to monitor the cafeteria during prep time, and she would not have time to call parents and consult with school counselors about difficult students. The cuts, then, would directly impinge on her ability to participate in the "professional obligations which make up the role of the teacher."[18] On September 22, the *Bulletin* published a letter from a teacher named A. Beaumont. She/he pointed out that the city asked teachers to "make the sacrifice" after giving raises to the city's police and firefighters without increasing their workload. Beaumont also excoriated Rizzo for pushing teachers to take cuts, "which shows where education comes on his list of priorities"; although she/he did not say it outright, Beaumont clearly believed that Rizzo's history as police chief and proponent of "law and order" undergirded his larger political strategy.[19]

Liberals had a different view of the strike. They attributed the largest share of blame to the school board and city leadership—Rizzo—for failing to provide the necessary resources. Highlighting the fracturing of the New Deal coalition, most members of this constituency were, like Rizzo and his supporters, likely to be Democrats, but they did not adhere to "law and order." Their critique of the strike focused on the importance of top-notch, state-financed public education in solving social and economic conflict through equality of opportunity. In their eyes, the city needed new revenue. Claire Kuperman, writing to the *Bulletin* on September 20, outlined her frustration about the strike. A parent, she believed that "many of the teachers' points are valid," but she also empathized with "the Board of Education which have been reduced to mere beggars, going from one government official to another, with hands out. We want a long-range solution to these problems, once and for all."[20] Jack Morris, writing to PFT president Sullivan the next day, suggested—instead of job cuts—a "campaign from the grass roots level" for more funds from the state and federal government. He asked that Sullivan use the picket lines to "get a flood of requests for financial help to Harrisburg and Washington."[21]

Two letters to the editor of the *Bulletin* on September 22 critiqued both metropolitan and national political economy. Commenting on an aid package that President Nixon recently promised Mayor Rizzo, Isabella Posch believed the city's priorities were out of line, hinting that gender differences might represent part of the problem: "With rage and shock, I just heard a report on TV that Mayor Rizzo is going to hire 900 more policemen. The money promised from President Nixon to meet Philadelphia's growing urban problems is most certainly welcome. However, first priority should be given to the educational crisis. If the mayor applied the same yardstick which he has applied to the striking teachers, he would have asked Philadelphia's policemen to be more dedicated and work longer shifts at the same salary." Jay Schuchar went even further, criticizing federal spending priorities. He pointed out that 250,000 Philadelphia schoolchildren missed school because of a $50 million shortage while the yearly appropriation for Vietnam was between $2.5 and $5 billion. He demanded the United States "stop bombing, burning, and maiming little children running naked and screaming with napalm burns from their schools, and spend the money to put our children back in their schools."[22]

Mrs. Peggy Definis, who lived in the white working-class northeastern section of the city, proposed a solution closer to home. She acknowledged that she "hate[d] paying taxes as much as anyone else," but she also recognized that Philly—because of the challenge of servicing so many students—"need[ed] to make salaries for teachers attractive enough that they'll put up with overcrowded conditions in the existing classrooms." She advocated a tax hike: "I think our children need and deserve an education, and I don't think I'm alone in saying I'm willing to pay for it."[23]

As late as December, liberal constituencies vehemently argued that more funding represented the key to solving the crisis and averting the second strike. A petition signed by thirty parents from a Northeast Philadelphia school appealed to Sullivan not to strike again: "Our children are the future of this city, and must receive a quality education. . . . There is no reason why it should not continue to do so with the cooperation of all concerned." The appeal assumed, however, that more money was needed. The school's parents had "worked hard to write letters, send telegrams, etc., to try and secure the necessary funds to keep our schools open through every avenue available to us." The union should "meet and negotiate in good faith and make a realistic settlement for the sake of our children."[24]

The president of the Finletter HSA, located in the multiracial neighborhood of Olney, mailed a similar, if more militant, letter to the PFT. Sending a carbon copy to the "editor" (the letter did not specify the newspaper[s]), Mrs. Joan Goldberg staunchly asserted that "we will not tolerate a second teachers' strike." The HSA, she pointed out, had "been in communication by means of meetings, telephone calls, telegrams, letters, petitions, etc., with state, city and national officials including the governor, the mayor, the councilmen, among others, arguing for funding of the schools." Now they had planned a mass rally as a means of "voicing our concerns to the elected officials and the public at large over this current school crisis."[25] Although this letter asserted the parents' frustration, it nonetheless pointed to the necessity of more funds as the ultimate solution to the crisis. A powerful critique of the existing metropolitan political economy was evident in the "liberal" discourse found in the public debate. Individuals and organizations argued that the city's and the nation's spending priorities were askew. Indeed, many Philadelphians clearly continued to endeavor to secure more funding for a public education system they believed was worth spending money on.

The third constituency, however, squarely blamed teachers for the schools' budget deficit. Following Rizzo's antitax, law-and-order politics, they believed the union should make major concessions and that austerity was necessary. Some of this group included families whose children attended parochial schools and resented paying taxes to an education system they did not use. Indeed, by 1970, Philadelphia's Catholic school enrollment—about 40 percent of the school-age population—was larger than any city in the United States except Chicago.[26] Still, most of this constituency did not publicly make the case that they should not have to pay for schools their children did not attend; their critique, rather, focused in large part on the supposedly excessive demands of teachers. Rizzo himself best summed up this position when, at the end of the September strike, he linked a tough, masculine stand symbolic of law and order with the "outrageous demands" of the union: "The union mistook kindness for weakness. If they think we're a bunch of patsies they're in for a rude awakening." He insisted that if he surrendered, "tomorrow every teacher will want a donkey."[27]

Critics built on this rhetoric to paint a picture of a teaching force that not only made excessive demands, but also produced little of value and took advantage of the "productive" citizens who paid the bills through their taxes. About two weeks into the September strike, for instance, Bill DeSimone wrote a letter to Sullivan in which he accused the union of "not being rational." He excoriated the PFT for its "unwillingness to accept a minimal increase in pay so that the budget may be eventually balanced," asking, "How much should teachers make? They only work for approximately 9 months a year." Further, DeSimone did not call himself a parent—though he was one—but instead referred to himself as a "taxpayer," asking, "How can we, as taxpayers, champion your cause?" He then provided a clue as to why he identified himself that way: he felt squeezed by economic conditions. Commenting on the teachers' resistance to a longer workday, he clearly resented their schedule: "How many poor low-level management people (most earning less money than your 'dedicated' teachers) spend 2, 3, and 4 hours at night after working a full day in preparation for the next day. You'd be surprised! Can't you see how ridiculous you sound whenever you try to defend your union. You're much better off than you deserve to be."[28] For "a poor low-level management" type like DeSimone, the crisis brought about by the strike provided an outlet to vent the frustrations of his own economic position, and it forced him to confront his relationship to the state. Indeed, no union fought for his economic rights, and he believed no organization advocated for him. The state, to DeSimone, was failing, as he believed it did little more than forcibly transfer his income through taxes to teachers who worked less than he did.

Others explicitly blamed teachers for the budget shortfall, linking, in their eyes, unproductive teachers and unproductive students. A Northeast Philadelphian writing to Sullivan, J. L. Simonsen, asserted that teachers should "cooperate" with the board of education "so we don't have to hear that the schools cannot remain open every year." Arnold Ifill wrote his newspaper that teachers were "asking for top dollar and producing some of the poorest quality of educated students in the country. When the students throughout Philadelphia start achieving average or better scores on the national tests, then the teachers should get a raise." Not only should teachers not get raises, but the board needed to "take appropriate action" against underachieving teachers.[29]

A key part of the antistrike, antitax discourse included an attack on a disproportionately female workforce for acting like other private-sector, mostly male unions. E. E. Pearson, for example, believed the strike represented the "harvest" of the "disastrous day in the field of public education when the AFL/CIO decided to organize the teachers." The teachers had adopted the "classic techniques of longshoremen, assembly-line workers, etc." who "with little provocation . . . walk off the job." Charlotte Paulsen wrote to the *Inquirer* from Ocean City, New Jersey, that teachers should not act like other workers. Invoking assumptions about the appropriate behavior of women charged with caring for children, Paulsen remarked: "If Philadelphia teachers are more concerned about a few dollars pay increase or a few extra hours work than their responsibilities to their pupils, they are not worthy of the status they presume as members of the teaching profession."[30]

In November, parents at the Stephen Decatur School—based in Northeastern Philadelphia—met to stop the January strike. In the meeting, HSA leaders distributed polls to parents to determine what to do. The ten polls sent to PFT president Sullivan that remain in the archive are invaluable as a window into the mind-set of the antitax constituency of white working-class Philadelphia. The first three questions asked participants if they would support efforts to meet with local council representatives, organize a bus caravan of parents to Harrisburg to pressure the state for more education funding, or to convoke a rally downtown. On these three questions, virtually every participant answered yes. The fourth question, however, asked if parents would "support a tax increase." With one exception, all parents answered in the negative, and even the one affirmative answer was heavily qualified.

What is even more illustrative, however, are the comments parents wrote in underneath the poll. An unnamed respondent made an argument similar to Jay Schuchar's, who had demanded that money allocated for Vietnam instead go to education: "Taxes are high enough. Use some of the money now being wasted in defense, civil service, etc. to pay for education—esp. federal tax money." A second respondent, Mrs. Margaret Griffin, agreed. "I feel we are taxed enough at present. If funds were utilized properly, additional taxes would not be necessary." And Mrs. A. J. Evans believed that "there is plenty of money around. It is used the wrong way and not for the right things."

Unlike the first respondent, Griffin and Evans did not specify what they meant when they said funds were used incorrectly. Perhaps they meant that the nation's priorities were out of whack or that Philly spent too much for police. Two additional respondents, however, made quite clear there was an additional interpretation. Mrs. Michael Sullivan, who supported a tax increase, did so only if "it included all sections of the city including a tax on apartment dwellers, because they also have children attending school." If the racial coding of "apartment dwellers" as opposed to the white property owners in Northeastern Philadelphia was not entirely explicit, J. E. Blasch made it unambiguous: "The Northeast pays enough taxes and gets little in return. Our taxes are supporting the ghetto areas."[31]

As the December deadline approached, the antitax constituency ratcheted up the pressure on the PFT. On November 28, a petition from thirty-five "tax paying parents" of students in a Northeastern Philadelphia school demanded that Sullivan "endeavor to secure funds from the City and State by Revenue Sharing or by some other means, but not by any more Taxes!" On December 6, Edward Schneider, also from Northeastern Philly, called the teachers' demands "insulting to the public at large. . . . The people are up to their necks in taxes and want no more and the urgent problem is now to get the School Board; City; and our Federal government out of debt deficits." Finally, Joseph Fayer, a center-city "father of four children and a taxpayer" threatened the union with the further erosion of the city tax base. "If there is another strike in January, I intend to move myself and my business out of Philadelphia. I feel certain many more middle class and upper class taxpayers will also move. In the long run your demands will result only in fewer teaching jobs and less money to run the schools."[32]

Living in the predominantly African American neighborhoods of West and North Philadelphia, the fourth constituency opposed the strike and castigated the union but for different reasons than the antitax group. This group believed Philadelphia children could succeed in a good school system. But, as had happened in New York City and Newark, some African Americans believed that the union—which was about two-thirds white—worked against reform. For reasons documented in chapter 2, Black Power advocates and the PFT had grappled in the late 1960s and early '70s, and some black Philadelphians believed that, at best, the union's efforts to maintain class sizes and improve teachers' salaries was not worth the time lost to the students. At worst, they argued, teachers may not have had the interests of their students in mind at all.

Mrs. Joy Brooks, writing to Sullivan from West Philadelphia in September, told him that "it is inconceivable to me, as a parent and a taxpayer that you can continue to force children to suffer a lack of educational advantage while you dicker over time and money. . . . You have the audacity to argue away our children's lives over 40 minutes and a few more dollars." Walter Desher, from North Philadelphia, condemned the union for its disingenuousness: "If you were really interested [in education] then you would make some compromises. . . . You have only stubbornly and bull headedly maintained the status quo." Another critic made a more cynical point. She asked if the children were "learning anything that will benefit them in later life" before the schools closed. She accused the board of wasting money, but also upbraided the teachers for striking: "Because the students are black," she argued, "we get a lot of bull-shit like always. If the system don't work get rid of it."[33]

Black students evidenced their own frustration with both the city's education system and the striking teachers. In a demonstration during the strike's first week, students protested at several government buildings, holding handmade signs that read, "Teachers Don't Give a Damn About Us. All they want is money, money, money"; "The 4 R's: Ross, Ryan, Rizzo, Racists"; and "Who are the Teachers Accountable to: the Students or the System?"[34] Many black teachers agreed with their students; the PFT's black caucus voted by a four-to-one margin against striking because they believed it served the interests of white teachers more than black teachers. According to one report, "most" of the 3,500 Philadelphia teachers crossing the picket line were black.[35]

As the second strike approached, Mr. and Mrs. Weber turned on their head what they presumed to be the teachers' racial assumptions about Philadelphia students: "To upset the normal learning processes and schedules which have been set up since the first stoppage would throw the entire system into a shambles. . . . PLEASE THINK OF THE CHILDREN! They're not all delinquents." Finally, Emily Crane from West Philadelphia urged Sullivan to "consider the unfairness of another strike. We all have to work for a living and teachers are no worse off than a good many other people." She suggested that, instead of striking, teachers "direct your talents and time to help those who want to develop some kind of on-going program to raise money for the schools each year rather than threaten the community with a strike each time a contract expires."[36]

As the second strike approached, these four constituencies vied to define the strike. As it dragged on through January and February, however, the insolubility of the labor conflict—highlighted by the illegal behavior of the teachers—emerged as the ultimate explanation for a dysfunctional school system. And the labor-liberal coalition continued to fracture. On January 3, the PFT membership, by a margin of almost five to one, voted to strike again, and the next day, the board of education held a televised meeting in which its members voted unanimously to seek an injunction to force teachers back to classrooms. The board's explanation for the action framed the city's financial dilemma as one in which teachers' demands—attenuated substantially from their demands the past September—necessitated a firm response. School board president Ross was in a tough spot: appointed to the board in 1965, Ross was a personal friend of Rizzo, who had elevated him to head the board. Ross also faced pressure from teachers and the city AFL-CIO and thus abstained from an eight-to-zero vote. Even so, he still rebuked the PFT, arguing that "compassion is not characteristic of our Teachers' Union in this City." They "want to brow-beat the whole community and of course, who gets hurt in the brow-beating, the children in the school system."[37] Ross, under pressure from the PFT, would end up resigning from the board before the strike ended.

Another board member—Dolores Olberholtzer—had no qualms about seeking an injunction. In fact, she questioned whether teachers should be allowed to collectively bargain at all because public-sector workers did not have to worry about losing business as did those in the private sector: "If a plumber messes up, they'll get called back. . . . If a teacher turns out to be a functional illiterate, that is just the breaks, you know, that's the way it goes." Even compared to other public employees, she reckoned, teachers' lack of productivity did not merit higher salaries. Citing only the time high school teachers spent in direct instruction, Olberholtzer argued that teachers only worked "a little over four hours a day." Not only did she support the injunction, but she also called on parents to assert their "right as a taxpayer" and "demand" that teachers go back to work.[38] Mayor Rizzo also held the line against the union on behalf of his constituency of "taxpayers." Commenting on the January strike, he asserted that he would not "capitulate to irresponsible demands" because a tax hike would cause "the people who are paying the tax load right now . . . [to] flee, leaving only those who cannot pay."[39]

Even during the early days of the second strike threat, the media, while supporting a tough stance against the teachers, nevertheless continued to ask bigger questions about how to solve the schools' fiscal crisis in addition to pressing Rizzo on his no-tax policy. An editorial reprinted from the newspaper the *Main Line Times* (published in the wealthy western suburbs) and endorsed by the *Inquirer* offered that the school system should be regionalized, asking, "When will we recognize the interdependence of the city and the suburbs . . . and start saving the whole region? If the city dies we will go with it, make no mistake." The *Bulletin*, while condemning the strike, also criticized the mayor for "ignor[ing] the legitimate and long overdue claim the schools have on local tax revenues which have yielded only an additional $30 million for schools in the past six years compared to a fourfold increase in state subsidies."[40]

Teachers, for their part, vocally defended the second strike and explained the budget deficit differently than the board and the mayor. H. H. McCreesh, a teacher whose epistolary defense was published in both of the city's newspapers, for instance, believed that "most parents, taxpayers, the general public, and editorial writers are unfairly critical of the Philadelphia Federation of Teachers and its membership." The financial crisis was "in no way the fault of teachers." He believed that "while a campaign of no new taxes is good politics, it is also very poor public policy."[41]

At the very beginning of the January strike, a lively discussion of how to solve the school system's financial crisis still existed. Two turning points during the course of the strike, however, minimized that part of the debate. First, on January 11, common pleas court judge Donald Jamiesen enjoined the strike, while the leadership and most of the rank and file refused to obey the injunction. On February 9, a jury convicted PFT president Sullivan and chief negotiator John Ryan of contempt of court and Jamiesen sentenced them to lengthy prison terms. Criminalizing the strike, these actions shifted the debate to one that focused almost exclusively on the teachers' tactics and, in the face of this intransigence, whether the city was capable of providing education at all; their "taxpayer" critics also rhetorically connected the PFT's lawbreaking behavior to an increasingly disaffected African American community, arguing that both groups in tandem caused the downward trajectory of Philadelphia.

Criminalizing the Strike

After almost four weeks of lost instruction time, Judge Jamiesen finally issued an injunction. Pennsylvania Act 195 provided state and municipal employees the right to strike, but a strike could be enjoined when it posed a "clear and present danger" to public welfare. Jamiesen's order argued that the strike presented a threat to the "educational welfare" of "low achievement pupils" as well as a threat to the "public safety" following "the release of children on the streets during the strike."[42] With regard to the latter assertion, one need not strain too much to see racial assumptions about law and order, particularly given the history of militant black student protest in Philadelphia in the late 1960s (see chapter 2).

The injunction highlighted the fundamental contradictions of PA 195: although teachers could strike, the "clear and present danger" clause could be defined at the discretion of the local courts. And, while a judge could fine and imprison both union leaders and even rank-and-file picketers, as the Philadelphia teachers showed, doing so was not likely to end a strike. Indeed, it is possible that the contradictory nature of strike rights for teachers in Pennsylvania gave school boards a deeply perverse incentive to stone-wall negotiations: they saved money during a strike when teachers missed paychecks, and if they waited long enough, they could get an injunction in the hopes of turning the public against teachers.

In fact, Jamiesen's injunction shifted the terms of debate in just that direction; instead of discussing the roots of Philly's financial crisis, critics focused on the illegality of the strike and increasingly blamed teachers for the school system's deficit. On January 12, for example, both the *Bulletin* and the *Inquirer* admonished teachers to obey the injunction. The *Inquirer* argued that "teachers who exercise their right to strike under the law also have a responsibility to respect the law's provisions regarding rights of the public." The *Bulletin* went even further, linking the union's intransigence to "its insistence on a settlement that far exceeds the school board's capacity to pay." The union could only "redeem itself in the public eye," according to the editorial, by "obeying the injunction and continuing the negotiation."[43] In a pattern played out elsewhere in the 1970s, the newspaper argued that teachers' financial sacrifice represented the only solution to a larger structural crisis.

Because the injunction defined the strike as illegal, it was not surprising to see Rizzo's followers question the assertive behavior of female teachers and to heighten the rhetoric of standing firm by focusing on upholding the "rule of law." Philadelphian Edward Jenkins, for example, wrote to the *Inquirer* accusing the teachers of "attempting to blackmail the two million tax payers of Philadelphia"; he hoped "Mayor Rizzo has the courage and fortitude to use all means to replace the striking teachers." M. Regina Black also wrote to the *Inquirer*, dividing the mayor's law-and-order constituency of taxpayers from a coalition of the lawless consisting of unproductive teachers and the students they taught: "Mayor Rizzo heads a large group of taxpayers who do not wish to pay more for substandard education of social delinquents who mar the daily existence of Philadelphia citizens with vandalism, graffiti, muggings, and public obscenities."[44] Other critics believed the teacher' illegal behavior threw gender roles into chaos. Eileen Antinnuci, for instance, used the example of a striking female teacher who called a male strikebreaker a "scab." Questioning the striker's proper role, she opined, "Here was an educated woman, a person supposed to mold young minds. She reduced herself to a cheap, loud-mouthed, sign-carrying law-breaker."[45]

The media also linked teachers, their students, and the decline of law and order in Philadelphia by calling for the court to prosecute the union leadership for contempt. On January 18, for example, WCAU-TV argued that "the teachers are taking the law into their own hands. We wonder how many times many of these same people have whined and cried about the lawlessness of their students. . . . What kind of teacher is it who teaches—by personal example—open defiance of the law?"[46]

A teacher named Winifred Rosenbaum had different ideas about gender roles than Eileen Antinnuci did. Rosenbaum believed the school crisis was "sexist." She argued that teachers should stay out of classrooms because the school board unfairly asked them to sacrifice. "Would anyone seriously suggest," she asked, "that policemen, firemen, lawyers, doctors, or any member of a male associated occupation make such noble sacrifices for the public good? Society is accustomed to making unrealistic demands on women, and to utilize them everywhere as cheap labor, that it is a natural carry-over to any occupation associated with them."[47]

The next two weeks brought the final escalation in the conflict. On Sunday, February 4, a four-hour televised debate between the board and PFT representatives indicating that a compromise was not forthcoming inspired a renewed round of anger from the antitax constituency. Even more important, on February 9, Sullivan and Ryan went to jail, and ten days later, the police arrested almost 400 picketing teachers.[48] Suddenly the "defiance" of the law was not merely an abstract question but a concrete development in which teachers were arrested, sentenced, and jailed.

On February 6, Mrs. John Boysen wrote to Ryan that she and her family had "watched the wrangle between the school board and teachers on Sunday. . . . I think you should know that none of the parents I polled are in support of your position." She believed that "the cost of our public employees is getting to the breaking point. You are asking the people of Philadelphia to support your demands from far poorer salaries than you command." And then, with a dire prediction indicating her fears of a racial hierarchy turned upside-down, she argued that "you are pricing yourselves right out of our lives. The city is emptying of its productive citizens who pay the bills. Soon there will be nothing left but white Catholics in parish schools, who have it tough enough already, and helpless blacks." The same day, an anonymous postcard, sent not to any of the PFT's male leaders but to Sunny Richman, a female negotiator featured prominently in the televised negotiations, unleashed the writer's misogynist rage at a woman who dared to lead a strike of schoolteachers: "A kid out of Temple U gets $8900 a year to start. That's

$234 a week for 9½ months work that's why the school system is broke!!! You Twat!!! You're nothing but a goddam thief!!!"[49]

The remaining two weeks of the strike featured vitriolic attacks not only on the PFT but also on the liberal state that had enabled the union. I. Buckman, for instance, believed the union "has already saturated their usefulness and value as a labor force in its short lifetime. True when they first organized, their salaries and benefits were inadequate, but through constant bargaining they were able to bring themselves to a very respectable level. But, as with many labor unions, their demands have become superfluous." Al Oakem, a retired teacher from Philadelphia, called the jailed union leaders "scofflaws" and hoped that "perhaps this country has at least reached the stage when it will move to protect itself from public servants who defy the law and cause irreparable social damage. Who knows, perhaps Philadelphia will become the site of a new declaration of independence—independence from unconscionable strikes by public employees."[50]

By the end of the strike in late February, an *Evening Bulletin* editorial showed just how far the meaning of the strike had shifted since its beginning in September 1972. The piece admitted the financial crisis had caused the strike and that the settlement had done nothing to solve it. More important, however, it argued that the primary lesson of the conflict was the failure of the state's attempt to solve the contentious problem of public-sector strikes: "Above all, perhaps, the school strike showed that Pennsylvania has still not found a reliable means of dealing with devastating strikes by public employees."[51]

Chastising the media for its naïveté, Theodore Robb criticized the fact that "all editorial writers seem convinced that the world's social ills can all be papered over with either dollars or new legislation." He believed that the only real solution to the strike crisis was for Mayor Rizzo to "ring down the curtain on Philadelphia." The strike also inspired the *Harrisburg Patriot*—recall the paper's optimism after the passage of the state's new public-sector labor policy in 1970—to reassess its view of the law: The ineffectiveness of the law in halting the Philadelphia strike "showed that Act 195 is in dire need of reexamination. Currently Pennsylvania and Hawaii are the only two states in the U.S. granting public employees even a limited right to strike. We think there should soon be only one: Hawaii."[52]

Much of the African American community was equally disenchanted. In a column for the city's African American newspaper—the *Philadelphia Tribune*—Pamala Haynes wrote a scathing rebuke of the teachers. Although she admitted that "it's not fair to say all of the teachers are a bunch of greedy, irresponsible persons who can't even show children how to read," she argued that neither the city nor the union valued education for nonwhite children: "The welfare of Black and Spanish speaking kids is no longer a priority, locally or nationally. Blacks and other people of color and the poor of all colors are now, once again, something to be controlled or pacified when possible."[53] Charles Bowser, head of the Philadelphia Urban Coalition and future mayoral candidate (in 1975 and 1979), believed that the financial deficit had only paralyzed the schools because the students were black: "If 60 percent of the students in public schools were white, there would not have been a teachers' strike."[54] Ali Robinson, in a letter to the *Tribune*, asserted simply that "we need very strong and valid reasons from a union leadership that refuses to teach our children especially when the union leadership is predominantly white and the students are predominantly Black. Has the union ever gone on strike for a more relevant education for Black children?"[55]

As both the white taxpayer constituency and the African American constituency pointed out, the school system did not appear to be working for anyone in Philadelphia—not for teachers, who feared chronic erosion of their pay from inflation and threats to their livelihood from fiscal crisis; not for students or their parents, who faced both teachers and a school system they believed had failed them; and not for the city's white working class, who believed they were victimized by the teachers' unfair demands and disrespect for the law and by paying for students who did not want to learn. A metropolitan—perhaps even federal—solution that radically reconfigured the power dynamics of the Philadelphia area would have been necessary to solve the problem in a way that would have ensured a stable school system and equal education for everyone. Unfortunately, the fracturing of the labor-liberal coalition had made such a possibility unlikely indeed.

Chicago: Third Strike in Five Years

When Philadelphia teachers geared up to strike in January 1973, so did AFT Local 1 in Chicago. Like Philly, Chicago faced a sizable budget deficit—just over $70 million; like the Philadelphia school board, the Chicago board refused to offer salary increases and wanted job cuts (1,200 out of the city's 26,000 teachers) and larger classes. The CTU initially demanded a 10 percent pay increase but quickly scaled it down to 2.5 percent; the salary increase was not as important as averting the layoffs and larger class sizes. As the strike vote loomed, the school board offered to punt the layoffs for six months while negotiations continued. Teachers voted 16,556 to 2,997 to strike.[56]

The strike began on January 10—two days after teachers in Philadelphia walked out—and lasted sixteen days, the longest in the city so far. Four days before the Chicago strike ended, St. Louis teachers struck. For several days, then, strikes simultaneously impacted the 950,000 students of Chicago, St. Louis, and Philadelphia.[57] The Chicago strike ended with teachers victorious, receiving the 2.5 percent pay increase and averting the layoffs. Still, as in Philly, the larger fiscal deficit remained unresolved: the *Chicago Tribune* pointed out at the end of the strike that the "$72 million deficit which led the board to say once that it had to retrench still stares everyone in the face," and both the CTU and the board hoped for "fresh infusions of money from the state's taxpayers" not likely to be forthcoming.[58]

In Chicago, the public discussion featured the same themes—in attenuated form—as those seen in Philadelphia. First, the demographic composition of Chicago—like Philly—opened up to conflict between the CTU and the black community. Most of the city's teachers walked the picket line; of the 1,300 crossing the line, however, a majority were African American. A CTU strike poll taken before the official vote indicated that a substantial number of black teachers believed "it is not in the best interest of black students and therefore cannot be supported."[59]

Some black community members agreed. As one letter to the *Chicago Daily Defender* put it: "It goes without saying that black students, who are in the majority of the city's public schools, always suffer." Echoing the voices of many African Americans in Philly, the critic argued, "If this were the other way around—white students in the majority—there would have been no teachers strike because the board would have easily solved its financial woes and strived to offer the best quality education in its 640 schools."[60] One black teacher summed up the state of events as teachers went back to work: "The strike has really split the black community and the rift may never be healed."[61]

Still, most black teachers honored the picket lines, and the most prominent civil rights organization in the city—the Reverend Jesse Jackson's Operation PUSH—supported the strike as well. Thomas Todd, PUSH's executive vice president, explained that if the board successfully cut the 1,200 jobs, black teachers—some of the last hired—would represent a disproportionate amount of the layoffs.[62] The *Daily Defender* also supported the strike. Perhaps because the CTU—as we have seen in chapter 2—had responded to the demands of black teachers, the black community overall held much less animosity toward the teacher union than in Philly. Indeed, the *Daily Defender* fingered a deficient system of funding education—rather than striking teachers—as the culprit in the Windy City. "The present impasse," the newspaper argued, "is borne out of the Teacher Union's insistence on conditions that would improve the teachers' lot. Their demands are neither excessive nor out of the scope of power of the Chicago Board of Education." Indeed, the editorial concluded that "the city of Chicago is not broke. If it gives the school system the kind of priority it deserves, money can be found to meet the teachers' salary scales."[63]

Other voices in the city, however, highlighted the fact that teachers could not legally strike in Illinois, and, like critics of the PFT, connected the city's failure to provide education to its inability to control its teacher workforce. The *Chicago Tribune*—the city's conservative newspaper of record—represented the loudest voice on law and order. The *Tribune* had long opposed the CTU; as early as 1965, when teachers were granted their initial representation election, the newspaper criticized the board of education's decision to allow it, asking, "Who is going to run the schools—the teachers or the duly elected or appointed representatives of the taxpayers? . . . If a school board is required to bargain each year with a teachers' union there will be an annual crisis, an annual threat of a strike, and possibly an annual strike."[64] As was the case in virtually every other state, Illinois law did not allow public-sector strikes (unlike New York, the law did not spell out stiff penalties for striking, however, leaving that to the judiciary), and the *Tribune* pointed that out before the strike began in 1973: "Strikes by public school teachers are illegal attempts to extort more money and power than teachers can get without resort to coercion." The piece concluded by calling on the board of education to seek an injunction against the teachers.[65] (The board declined to do so, and this restraint might explain why the Chicago strike was much shorter than that in Philadelphia.)

In an editorial on January 11 (the first edition after the strike began), the *Tribune* emphasized the city's victimization resulting from teachers' third strike in five years. Putting the strike into the recent history of successful teacher unionization, it believed that "the AFT will go as far as flabby, irresolute boards will let it go."[66] Two days, later, the newspaper linked Chicago to Philly, pointing gleefully to Judge Jamiesen's injunction and beseeching the Chicago board to seek the same punishment: "As each day passes Chicagoans and Illinois citizens become increasingly alienated from striking teachers and from public officials who knuckle under to insatiable demands and illegal tactics."[67]

Disappointed by what it viewed as the capitulation of the school board, and after the Philadelphia strike finally came to a close, the newspaper assessed the outcome of the strike in the City of Brotherly Love. The *Tribune* believed that Rizzo's search for new taxes to pay for the teachers' raises would represent a "long search." While it lauded the mayor's aggressive attempt to stand up to the Philadelphia teachers, the *Tribune* held out little hope for limiting the power of unions. It argued that Philadelphia showed that teacher unions could "lick" even "a determined school board. . . . If the AFT is ever going to lose a major confrontation, it will be in a city willing to pay an even higher price in disruption of essential services than Philadelphia paid. Is there such a city? It certainly is not Mayor Daley's Chicago."[68]

St. Louis: Illegal from the Start

The St. Louis strike that began while teachers in Chicago and Philadelphia were out also revolved around the city's political economy and the illegal behavior of striking teachers. Teachers organized the St. Louis Teachers Union (STLTU–AFT) in the 1930s, but it was not until the late 1960s that teachers began to successfully push toward collective bargaining. The state's 1959 public-employee law allowed teachers' organizations to "meet and confer" with school boards, but did not obligate boards to bargain or even meet with teachers. Public employees, of course, could not strike.[69]

As the result of "teacher militancy" in the Gateway City—in the words of STLTU president Demosthenes DuBose—the St. Louis Board of Education gave teachers a raise in 1969 but refused a representation election. For the next three years, teachers received no more raises, and inflation eroded their purchasing power. Indeed, teachers pointed out that, by 1972, a skilled union craftsman without a high school education working for the school district began with an annual salary that was $3,000 higher than a first-year teacher. Teachers believed that their level of education and social status entitled them to be paid on par with or better than blue-collar workers. Without exclusive representation, two unions competed for the city's teachers—the slightly larger STLTU (Local 420) and the St. Louis Teachers Association (SLTA–NEA). In December 1972, the two sides joined forces, threatening a strike the next January unless the school board agreed to a collective-bargaining election, hospitalization insurance, and a "substantial" midyear raise.[70]

The board not only refused to negotiate but also argued that even bargaining with teachers was illegal. Board members further claimed that although the teachers deserved raises, the St. Louis school system already ran a deficit of several million dollars and could not provide them, particularly in the middle of the year when the budget had already been fixed. Union leaders—especially DuBose—disputed the board's numbers, arguing that their assertion of deficit spending relied on holding millions in reserve as a hedge against declining revenues. The budgeting formula would become a key point of contention in the ensuing strike, but it was clear that, like Philadelphia and Chicago, St. Louis schools faced a systemic revenue problem as higher-income residents continued to move out of the city in the early 1970s and property values declined, thus leaving a lower property-tax base behind.[71]

When the union held a strike vote, St. Louisans could already connect the conflict to Chicago and Philadelphia. Local reportage—like that in the *St. Louis Post-Dispatch* on January 12, 1973—linked the Gateway City's first teacher strike to a nationwide "conflict over the issues of education and teachers' rights" that was "about to bear down upon the 4,170 teachers and 103,987 students in the St. Louis system."[72] Unlike in Chicago, local authorities were not reticent to try to halt the strike (in 1968, as documented in chapter 1, St. Louis teachers obeyed an antistrike injunction). St. Louis circuit court judge Thomas McGuire thus criminalized the strike from the beginning, issuing a restraining order to prevent strike leaders from calling on teachers to walk out. On January 22, however, 3,000 teachers went on strike anyway, and the number increased to 3,500 the next day. Five days later, McGuire issued a formal injunction, asserting that not only was it illegal but also, as the board had argued, that Missouri law made any collectively bargained contract illegal. By the middle of February, the judge had levied substantial fines on both the STLTU and the SLTA, and leaders of both unions disappeared from the public in order to avoid arrest.

Both the unions and the board called for more state aid from Republican governor Kit Bond, and, in yet another example of the problem with the multilevel structure of funding education in the United States, Bond promptly declined, calling the strike a "local matter." Of the school system's $86 million budget, around $30 million already came from the state, as each school district received funds based on the number of students attending school.[73] It was not until the St. Louis board of aldermen diverted $1 million in federal revenue-sharing money to the schools that the board of education agreed to a modest midyear raise and promised to increase salaries the next year as well. Teachers ultimately ended up with a raise of around $600 across the board from a lowest starting salary of about $7,200.[74] The board also agreed to pay for hospitalization insurance and to allow a vote to determine "majority status" for one single organization.[75] Though minimal, it represented a clear victory for the strikers, and, in spite of the substantial fines the two unions had to pay, they had forced the board of education to bargain. Further, the election making STLTU the collective-bargaining agent for St. Louis teachers in 1974 would simply not have been possible without the strike.

The narrative that played out in the media was multifaceted and somewhat different from that in Philadelphia and Chicago. Labor politics in St. Louis straddled the line between the liberal support for organized labor found in industrial states like New York and Pennsylvania and the Sunbelt labor politics defined by individual states' ability to pass "right-to-work" laws institutionalized in Taft-Hartley. With sizable unionized workforces in Kansas City and St. Louis, Missourians had rejected the right-to-work law that three of the Show Me State's four neighbors (Kansas, Iowa, and Arkansas) had on their books. Support for labor, however, clearly had not reached the same level whereby public-sector workers enjoyed rights similar to those in the Northeast, and the school board, unlike Chicago's, was hostile to collective bargaining.

In such a climate, it is not surprising that two seemingly contradictory narratives emerged simultaneously. According to some explanations, the crisis was caused by teachers who refused to abide by the law. In the eyes of others, an unfair funding system caused the problem: teachers deserved more money, but either the St. Louis government and/or the State of Missouri had not adequately funded city schools. The government's failure—both to control its labor force from illegal action and to provide the necessary resources for the citizens of the city—brought the two explanations together. Even the conservative *St. Louis Globe-Democrat*, for example, issued many patently antiunion editorials, but it also argued that Missouri had failed to adequately support the schools.

On January 19, for example, the *Globe-Democrat* explained the conflict as the result of "a crew of agitators" from the NEA and the AFT who were "stirring up strike trouble." Using language that echoed white Southern opposition to the civil rights movement, the newspaper asserted that these "outsiders" did not have the "best interests of the community at heart." The editorial lauded the court's decision to issue a restraining order. At the same time, it also admitted that "St. Louis teachers deserve the pay raises they are pushing for" but should wait until the school system moved beyond "its presently precarious financial position."[76]

Once the strike began, the teachers' "illegal" action took center stage. Clyde Miller, the city superintendent, wondered, "Will St. Louis public schools be surrendered to mob rule? If they are, then the destiny of the teachers, the children and the community will be cast into the hands of the strongest muscles at the moment." In an editorial on January 24, the *Globe-Democrat* echoed Miller's concerns. The newspaper called the striking teachers "heady with the wine of mass disruption" and charged them with "resort[ing] to mob action to close the schools still operating after the start of their strike." The editorial argued that "such hooliganism might be expected of some student radicals, but it is abhorrent coming from the Teamster-affiliated St. Louis Teachers Union or the more professional St. Louis Association of Teachers."[77]

KMOX—the city's premier news radio station—also excoriated the teachers. An editorial admitted that teachers needed better salaries and argued that "teachers, like other public employees, should have a right to organize and bargain collectively." Nevertheless, the piece banged the drum of law and order, lamenting the decline of respect for the law so many other witnesses attached to teacher strikes elsewhere from the mid-1960s on: "Our children today see enough examples of flaunting [sic] the law. Teachers fail their duty to the young if they themselves do not respect the laws of the land."[78]

When the restraining order became a formal injunction, the media's attacks deepened. The more sympathetic *Post-Dispatch*, on February 1, for example, believed the injunction represented a "logical legal development." The newspaper lectured the strikers on fiscal realities: "The Board's financial position seems not to have impressed the strike leaders, but we think it should." The piece called on the teachers to end the strike and work with the board to get new funds. The editor at the *Globe-Democrat* was apoplectic that teachers defied the judge's order: "At this point the teachers have about as much logic on their side as the fabled army officer in Vietnam who reportedly told his troops: 'we're going to have to destroy this town to save it.'"[79]

On February 6, Judge McGuire fined the SLTA the enormous sum of $150,000 for violating the injunction (SLTA president Jerry Abernathy had been more vocal in defying the injunction than the STLTU leaders). Just as it had in Philadelphia when strike leaders went to jail, the hefty punishment helped to reframe the teachers' actions at the center of the conflict. A resident of suburban Florissant, for example, was "disheartened to observe the defiance of a court order." He longed for the time when teachers had "imparted not only the basic knowledge we had to possess but more than that, a deep moral sense of responsibility, respect for authority, courtesy, and love of home and country."[80] Such a call for the good old days could not be separated from the notion that most of those teachers exhibiting obedience and inculcating morals were women who toiled at lower salaries than male blue-collar workers.

Unlike in Philadelphia, however, the strike did not last long enough nor were the punishments severe enough for the discourse of the criminal teacher to totally eclipse the discussion of the intractable problem of metropolitan political economy or the failure of the state to solve it. St. Louis, like many industrial cities in the United States, faced deindustrialization almost at the moment the nation began demobilizing from World War II. The city followed a similar trajectory as many of the other cities we have seen so far: African Americans—many from the rural South—continued to migrate to St. Louis in the years immediately after the war while the jobs moved—along with white single-family homeowners abetted by federal housing policy—to the suburbs or to the more capital-friendly environs of the Sunbelt.[81]

The population of the city, in fact, declined from a high of about 850,000 in 1950 to around 620,000 in 1970; 517,000 in 1977; and 453,000 by 1980. Much of this decline represented the continuing movement of white property-owning families into the suburbs; in 1970, the African American population of the city made up about 40 percent of the total population (a dramatic increase from 13 percent in 1940), but the percentage of blacks in the public schools stood at more than 70 percent. These demographic changes led to a substantial drop in the property-tax revenue needed to finance the schools; not only were there simply fewer taxpaying citizens, but in 1970, one-third of the city's residents were on welfare rolls—thus not likely to contribute much in property taxes. St. Louis's secession from the county in the nineteenth century also permanently fixed the city's borders, so it could not add any more taxable communities.[82] The school system, as a result, relied on major injections of revenue from the State of Missouri to meet its financial obligations. Therefore, the only options for teachers to secure raises would be an increase in property-tax rates, more state aid, or deficit spending.

The board unequivocally refused to finance raises through debt, so public discussion focused on the other two alternatives—raising tax rates or getting more money from the state—and whether they viewed the strike as an unwelcome necessity or an immoral action, commentators criticized the failure of the government to solve the problem. The *Globe-Democrat*, for example, augmented its antiunion position with one that called for more aid from Missouri. The newspaper's editorial in its January 27–28 weekend edition pointed squarely to the city's declining property-tax base and underscored the state's unfair formula for aid to districts: "Presently, state-wide funding averages 42 per cent. But St. Louis, which has a greater need than most school districts, gets only 37 per cent! . . . It should be obvious that a maximum effort must be made by the city and everyone interested in better schools to gain passage of [a new] legislative package. The future of St. Louis schools is at stake. And what could be more important to a city?"[83] The *Post-Dispatch* argued a similar point. Although the editor criticized the "total irresponsibility" of the teachers' "illegal strong arm tactic"—like Rizzo's language in Philadelphia, the newspaper editor's language hinted at the inversion of gender power—he also believed the school system "must have major assistance from the state to avert a crisis in the years immediately ahead."[84]

Letters to the editor also blamed the strike on a failed system. Many of these came from teachers like Bill Diffley, who asserted that "dedication" would not get him a discount when he paid his bills.[85] Others, however, who were not teachers, also criticized the system of American metropolitan political economy that seemed to force teachers to strike. From across the Mississippi River in Edwardsville, Illinois, Nancy Duncan argued that "until the parent cares as much about his child's education as he expects the teacher to care, his child's education will be deficient. . . . The taxpayer turns down school tax appropriations because he can. He pays his military taxes because he must. The former results in larger classes with less individual and remedial student learning. The latter pays for millions of dollars in cost over-runs and misguided wars. Isn't our country's best defense our children, not bullets?[86]

Nor were these critiques written out of the explanation for the strike—as they were in Philadelphia—when teachers were punished. For example, a student from just outside the city—in nearby Clayton, Missouri—wrote to the *Post-Dispatch* several days after McGuire fined the teachers. He believed the strike highlighted the inequity of the metropolitan school system: "As a student at Clayton High School, I can find no justice in the fact that school is continuing as usual in Clayton, while just across Skinker Boulevard [the boundary between St. Louis and Clayton] my fellow students in the city are being deprived of their education through no fault of their own. The teachers' strike is, however, merely a manifestation of a much greater national problem: the inequality of our school systems." He concluded that "huge sums of money should be appropriated to raise other districts to the level of Clayton."[87] Several days later, Francis Thomas characterized, more vociferously, the strike in similar terms: "The fundamental cause of the St. Louis school strike is not the militancy of the Board of Education but the niggardly St. Louis taxpayer." In particular, he castigated white middle-class taxpayers in St. Louis:

> My children attend parochial schools, so why should I pay for public schools? The children in the schools are predominantly black, so why should I be concerned about them? These are some of the arguments advanced by many who pay lip service to education but are unwilling to pay for a first-rate public school system. The school tax rate in St. Louis is far lower than that of most St. Louis

County districts; until St. Louisans can be persuaded to adopt a more altruistic attitude toward the financing of their public school system it will continue to be troubled by the financial uncertainty and justifiably dissatisfied teachers.[88]

In the waning days of the strike, even the *Post-Dispatch*, which had called on the teachers multiple times to go back to work, asserted that the most important question to emerge from the strike resulted from a subpar system of financing. On February 16, an editorial urged Governor Bond to find a way to bring more money to St. Louis. The piece urged Bond to consider the example of Detroit, then $90 million in debt and considering closing the schools before the academic year ended: "If something is not done fairly soon" in St. Louis, the editor urged, "the situation could continue to deteriorate, with disastrous consequences."[89]

On May 10, 1973, the *Globe-Democrat* offered a concluding assessment of the St. Louis strike. Consistent with its vehement antistrike position, and just as the *Chicago Tribune* had done, it accused teachers of having "bludgeoned the Board into granting a [salary] increase." At the same time, the way it characterized the board's efforts is instructive. Although the teachers had extorted the "public" by illegally walking off their jobs, the editor reserved most of his disapprobation for the board. Indeed, the board had "knuckle[ed] under" to the teachers. Now, it needed more funds to pay higher salaries and was, according to the *Globe-Democrat*, left with a lose-lose choice between deficit financing or trying to convince taxpayers to approve higher property taxes.[90]

Conclusion and Postscript: Detroit

Unlike in Philadelphia, where the length of the strike, the city's strike history, and the harsh punishments meted out to teachers erased a discussion of the city's political economy, that discussion continued to occupy a prominent place in St. Louis until the end of the strike. Ultimately, however, critics emphasized that the labor-liberal state had failed in St. Louis just as it had in Philadelphia. The city government could neither control its own labor force in St. Louis by using the legal system to get teachers back in schools nor could it (in conjunction with the State of Missouri) adequately solve the problem of financing the school system. The evidence of the backlash against the city's "unproductive" also had not yet really emerged in St. Louis as it had in Philadelphia. Still, it is clear that the lengthy strike in the Gateway City—as it had in Philly—represented a space in which many people began to rethink their political assumptions.

In 1979 and 1981, respectively, teachers in both St. Louis and Philadelphia undertook another long and controversial strike. By then, however, the political climate in the nation had turned much more pessimistic. As we shall see, the fiscal limitations became an even more significant agent in reformulating political allegiances in each city.

A seven-week teacher strike in Detroit in September 1973 contentiously reprised conflict similar to that in Philly, Chicago, and St. Louis earlier in the year. Jeffrey Mirel has treated the conflict in his account of the decline of the Motor City's school system, and he argues that larger political economic forces were to blame in Detroit—in addition to both the school board and the teachers—for a divisive strike that solved very little.

By the early 1970s, a series of problems plagued funding for Detroit schools: declining property values and white working-class taxpayer resistance against higher levies, increasing student enrollments, and teachers' demands for higher salaries. The city faced massive budget deficits (about $75 million in 1972) and teachers had forgone raises in 1971 and 1972. After the legislature provided the central board of education with the power to impose a higher property-tax millage, teachers asked for an 11 percent pay increase—the equivalent of inflation over the past two years.[91] The school board would only agree to raises if teachers agreed to a new performance-accountability plan that threatened teacher tenure. For many of the same reasons the UFT shut down education in New York over the reassignment of a handful of teachers in Brooklyn, Detroit teachers went to the mat for job security. As in Philadelphia and St. Louis, a local judge enjoined the strike and held the DFT in contempt of court, punishing the union with substantial fines. After forty-three days on strike, the teachers returned to classrooms—agreeing to mediation on salaries, while the governor punted the accountability issue by setting up a statewide panel to study it. As Mirel concludes, "perhaps the only thing the strike accomplished was to reveal still further the fragmentation of the liberal-labor-black coalition."[92]

It should be added that the public discussion in Detroit also revolved around the illegality of teachers' actions. Although strikes were against the law in Michigan, it was rare for local judges to issue injunctions—in fact, the Detroit strike of 1967 had not been enjoined. Judge Thomas Foley, however, broke precedent in 1973. Although his decision admitted that school board negotiators had "failed to recognize their legal obligation to bargain in good faith," he ordered an injunction anyway, on account of the importance of "the welfare and paramount interests of some 275,000 children and young adults and, further, the continuance of a viable Detroit education system."[93]

Further, the prism for understanding the teachers' actions—because the DFT was one of the few big-city locals represented by a female president—prominently featured gender. Much of the public discussion, in fact, focused on Mary Riordan and her defiant assertion that she would "go to jail" rather than call off the strike. The *Detroit Free Press* called her stance an "act of foolishness" and demanded local authorities use "the full force of the law" to end the strike.[94] After the court levied stiff fines—$1,000 a day for the individual leaders of the DFT and $11,000 a day for the union—the same newspaper featured a political cartoon in the tradition of the "frenzied feminine finance" critique of Margaret Haley in the 1910s. In the cartoon, an animated woman who looked strikingly like Riordan (with gray hair and glasses) held a placard displaying a dollar sign. A masculine hand holding a paddle emblazoned with the words "Tough Court Action" put the DFT president back in her place.[95] Germaine Leavitt, writing to the *Detroit News*, wanted to take Riordan up on her offer to go to jail: "I find myself completely lacking sympathy" for the DFT and "particularly" for Riordan. Indeed, Leavitt wanted to be sure that the female leader of the union was held to the same standards as male unionists: "When a person takes the law into his own hands and disobeys it, he should be jailed. What would result in this city if firemen and policemen were to strike?"[96]

As in Philly, the Detroit media also asked for a new statute to ensure teachers could no longer strike. On October 4, 1973, for example, the *Free Press* called on the legislature to "put precise and tough penalties into the [public-employee relations] law, and . . . rewrite it to give judges clear authority to order schools reopened and public employees back to work when the work stoppage is illegal."[97] How such a law would have solved a conflict like that in Detroit—when teachers already vowed to go to jail—was unclear.

Going forward, however, it was evident that many in the Detroit area felt like Joe Stroud, an editor for the *Free Press*. In a piece published on October 3, 1973, he pointed out how his father had moved *to* Detroit to seek an education. By 1972, he admitted, as the school system faced a massive deficit, he and his wife enrolled their high school freshman in a private school. Although he argued that he still believed in public schools, he would not let his children "become victims" of "Detroit's difficulties in providing a decent education system." After the district received the authority to raise taxes, Stroud had felt optimistic, but then Detroit teachers engaged in "the biggest, the longest, and the most insane" strike of the fall. "It is almost too much," the editor complained. "Detroit at such times seems incapable of self-government. The board and the union seem incapable of responsible bargaining on any consistent basis. . . . In any case, one is left to wonder whether the forces are too big, too bitter, and too fouled up to be managed." Indeed, after the strikes in each of these cities, one gets the sense that many of their citizens had begun to feel like Stroud. From his class position as a newspaper editor, however, he enjoyed options that many of these people—white or black—did not have. For them, they were just left with the bitterness—and the feeling that the political coalition that had done so much to make their lives better was unable to do much for them going forward.

CHAPTER 4

Dropping Dead

Teachers, the New York City Fiscal Crisis, and Austerity

On October 29, 1975, President Gerald Ford gave the speech that led to the iconic front page of the next day's *New York Daily News* proclaiming "FORD TO CITY: DROP DEAD." President Ford never uttered such dire words, but the line pretty well summed up his speech to the National Press Club that night. Further, his message was far from surprising; that summer and fall, Ford criticized Gotham's supposed profligacy as part of a campaign to parry a primary challenge on his right by former California governor Ronald Reagan. As a political calculation, then, Ford's response to New York that fall tells us much about the flux of political ideas in the mid-1970s.

"Some contend," Ford began, that the crisis resulted from "long-range economic factors such as the flight to the suburbs of the city's more affluent citizens, the migration to the city of poorer people or the departure of industry." The "straight talk," the president argued, was that "most other cities in America have faced these very same challenges, and they are still financially healthy today. They have not been luckier than New York; they simply have been better managed." The sins committed in the city's recent history, to Ford, stemmed from capitulation to freeloaders:

> The record shows that New York City's wages and salaries are the highest in the United States. . . .
> The record shows that in most cities, municipal employees have to pay 50 per cent or more of the cost of their pensions. New York City is the only major city in the country that picks up the entire burden.
> The record shows that when New York's municipal employees retire they often retire much earlier than in most cities and at pensions considerably higher than sound retirement plans permit. . . .
> The record shows New York City operates one of the largest universities in the world, free of tuition for any high school graduate, rich or poor, who wants to attend.

As for New York's much-discussed welfare burden, the record shows more than one current welfare recipient in 10 may be legally ineligible for welfare assistance.[1]

Ford's speech turned what had been a robust discussion in the city's public sphere of the structural impediments to the promise of postwar labor liberalism on its head. In the president's mind, New York's problem had been fulfilling the promise of the New Deal, and later, Roosevelt's World War II vision of "freedom from want": paying its workers a living wage and securing their retirement, providing medical care to its citizens, providing equality of opportunity to its youth through free college tuition, and ensuring welfare benefits for those who could not work.

As the president, Ford's may have been the loudest national voice to make this argument, but his was neither the first nor the last. Indeed, in New York, the city's yearlong specter of default led to a cacophonous debate over what had caused its fiscal shortfall and who should pay the price as life in the city was restructured. These debates—which took place on both the local and national level—led to a struggle to remake the city fiscally, politically, and ideologically, and, in turn, to limit the possibilities for social democracy in the United States. Indeed, whereas early in the crisis many voices attempted to revivify postwar liberalism in order to alleviate the structural difficulties faced by many American cities, a year of bare-knuckle contests played out in the media pushed less complex explanations that revolved around defining the limits of the "productive" citizenry. In an economic climate in which high unemployment rates and inflation caused pain for New Yorkers, public conversations fostered a view in which many of those who identified themselves as middle class fixated on blaming the "unproductive"—including both public-sector workers and welfare recipients—for the fiscal crisis. As seen in Philadelphia, this new narrative limited the options for solving the crisis and undermined the labor liberalism that had structured city politics for a generation.

In spite of its importance in both American urban history and national politics, the New York City fiscal crisis has received little in-depth treatment by historians. The best scholarly treatment from a historical perspective is that of Joshua Freeman, whose chapter on the crisis in *Working-Class New York: Life and Labor since World War II* shows that 1975 was so important in the nation's history because by that point New York represented the exemplar of what the postwar labor-liberal alliance had accomplished. The supposed profligacies pointed to by Ford—generous pensions, a free university education, robust welfare system—represent, to Freeman, working-class New York's "historic achievements," and the importance of the fiscal crisis was to undo them, both in the city and elsewhere. "Because New York served as the standard-bearer for urban liberalism and the idea of a welfare state," writes Freeman, "the attacks on its municipal services and their decline helped pave the way for the national conservative hegemony of the 1980s and 1990s."[2]

This chapter shows that the debate in New York City and in Washington over who was to blame for the crisis shifted not only policies but also the assumptions behind future policies during the course of 1975. For much of that year, loud voices repeatedly made the case that the city's plight stemmed not from the excessive demands of unions or from wasteful spending but because New York City faced a heightened version of the complex difficulties of many American cities in the late 1960s and '70s. Everyone from Mayor Abraham Beame to Albert Shanker tried to advance the plausible argument that the longer-term trends of deindustrialization, suburbanization, and an idiosyncratically high welfare burden left unattended by the federal government had caused the city's unprecedented money crisis. Ultimately, however, as in Philadelphia, the banks pushed hard for discipline. With little interest from either the State of New York or the federal government in alleviating these structural problems, the Beame administration could only capitulate to demands for massive cuts to the city's workforce, pitting the mayor directly against New York's public employees. In turn, the latter were forced to use aggressive tactics like strikes, sick-outs, and protests to mitigate some of the worst of these cuts, and many of these actions were illegal. Even this level of pushback, however, was not enough, and finance interests largely ended up dictating the terms through which the city would be restructured.

Whatever successes public-sector unions accomplished came at the expense of further empowering opponents to argue that they were not interested in saving the city but only illegitimately continuing to line their own pockets. Even though municipal-worker pension funds accounted for a larger investment in the city's debt than the federal government's, the conventional explanation for the crisis by its end centered on the intransigence of the unionized workforce, discursively linked to welfare recipients who many argued were unwilling to work. The city's teachers—in part a result of their long history with divisive labor conflict—represented arguably the most prominent group of city workers to take part in this battle. Many critics linked teachers with male, blue-collar public employees like sanitation workers, police officers, and firefighters, and the most prominent symbol of teachers' excess was Albert Shanker. Nevertheless, given that critics linked teachers most closely to welfare recipients—a population also clearly gendered—it seems likely that a good deal of the rhetoric around teachers stemmed from the gendered nature of their work.

Teacher Strikes: 1974

The Detroit strike in the fall of 1973 examined in the previous chapter represented the longest urban teacher strike of the 1973–74 school year. The spring of 1974, however, brought other lengthy and divisive strikes that heightened the national narrative around the battle zone that urban education had become. In February, for instance, teachers in Baltimore went on strike for a month. As in St. Louis, it was a joint strike: the Public School Teachers Association (PSTA–NEA) and the Baltimore Teachers Union (BTU–AFT) sought wage increases of 2 percent for the remainder of the 1973–74 school year and an 11 percent package of salary and fringe benefits for 1974–75. Union leaders pointed out that inflation had caused consumer prices to increase 13 percent since their last pay increase.[3]

Baltimore, like Philadelphia and Detroit, suffered from a fiscal crunch in the 1970s. Like St. Louis, Baltimore had seceded from the surrounding county in the nineteenth century and so could not annex any new taxable areas. Also, as in St. Louis, the public discussion of the monthlong strike in Baltimore focused on obtaining more state aid.[4] Although the strike was illegal, the mayor of Baltimore, Democrat William Donald Schaefer, responded by asking the judge presiding over the Baltimore Board of Education's injunction request not to punish the teachers.[5] Ultimately, union leaders decided against drawing the strike out any longer when Governor Marvin Mandel (D)—reluctant to raise taxes in an election year—refused to add any significant revenue to support education in Baltimore. Instead, teachers settled for the board's offer of a minimal raise well below the rate of inflation.

Though no public narrative of teacher culpability developed in Baltimore in February 1974, the lengthy school shutdown weighed on the nationally controversial sanitation and police strikes that exploded in the city the following July, and discussions centering on the illegality of public-sector strikes in Charm City mirrored those in Philly in 1973. For the same reason teachers believed it necessary to strike against the cash-strapped city in February, so Baltimore's sanitation workers—beginning in a July 1 wildcat strike—and then a large portion of the city's police from July 11–15 struck to gain higher salaries.[6] Although the city exacted retribution—by laying off workers to pay for wage increases and firing eighty probationary policemen—these two strikes succeeded, gaining sizable pay increases.

The two municipal strikes—especially the police strike—galvanized a local discourse in which the city's unions caused runaway taxation and a breakdown in law and order. Two days into the police strike, for example, Baltimorean Adeline Berluti connected the actions of public-sector workers to the city's tax base: "I deplore and condemn strikes which jeopardize the welfare of the general public. Such strikes should be not only illegal but absolutely not tolerated. Civil workers get paid thru taxation and this taxation falls on only a segment of the population, that is, the property owner." "Three voters" in another Baltimore household connected the police strike directly to the February teacher strike when they argued that "all city workers (including teachers) elect to work for the city of Baltimore . . . since the budget was already approved, the city workers currently on strike are breaking the law." They urged the mayor to "stick to your guns. . . . If you see fit to call the National Guard to help police the streets and empty the garbage, we are sure you will have more support from the tax-paying public than you have ever had from your own city employees."[7]

In March, teachers walked out of their classrooms in San Francisco and Kansas City. San Francisco's 2,500 AFT-affiliated teachers went on strike on March 7 demanding a 15 percent raise, while the school board offered 6 percent. The 2,000 teachers of the NEA affiliate—the Classroom Teachers' Association—joined the AFT contingent too, but after three weeks, both groups returned to work with nothing more than the original offer.[8] Teachers in Kansas City walked out on March 18 seeking a substantial pay increase, and the success of St. Louis teachers in early 1973 provided an effective example of collective teacher power in Missouri. After the district closed the schools a week later, the AFT-affiliated teachers stayed out another month, the union was fined $50,000, and Local 691 president Norman Hudson served eight days in jail. The teachers won a hard-fought 8 percent pay increase, however, with the promise of an additional 2 percent if a tax levy on the ballot that summer passed. On May 2, 1974, *NBC Nightly News* featured Kansas City, and the reporter on the scene called the strike "traumatic" to the community.[9]

Two small-town strikes in 1974 also garnered national media attention. First, about one hundred teachers in the small Timberlane School District in New Hampshire walked out in February. The school board fired and replaced them, but teachers still picketed the schools as late as September 1974.[10] Two weeks after the strike in the Granite State, teachers in the tiny rural Wisconsin district of Hortonville went on strike. During the course of a bruising conflict that divided the town and the state of Wisconsin, the Hortonville board fired and replaced eighty-four teachers. The teachers appealed—unsuccessfully—to the U.S. Supreme Court.[11]

The series of teacher strikes in early 1974 fed into a national narrative of a seemingly interminable strike wave (prominently including the Baltimore police and sanitation strikes) in the early summer of 1974. Total strikes by all workers in groups of 1,000 or more spiked from 250 in 1972 to 424 in 1974, and the number of workers on strike nearly doubled in the same period, from 975,000 to around 1.8 million.[12] Contracts timed to expire with the Nixon administration's inflation controls led to strikes in many different parts of the nation. As the sanitation and police strikes concluded in Baltimore, for example, state employees in Ohio's prisons and mental institutions, machinists for National Airlines, and San Francisco transit workers were also out on strike. Syndicated stories with headlines like "Walkouts Disrupting U.S. Cities" and "Nearly 600 Strikes Grip Nation" proliferated in American newspapers.[13]

Then, in the fall of 1974, the AFL-CIO convened its new Department of Public Employees, which included the AFT, AFSCME, the American Federation of Government Employees, the American Postal Workers Union, the Transport Workers Union, and the International Association of Fire Fighters, among others. AFL-CIO president George Meany gave a fiery address at the convention asserting that "I don't think it makes a great deal of difference when you get down to brass tacks whether there's a law which affirmatively says you have a right to strike. . . . As far as I'm concerned if you treat public employees bad enough, they'll go on strike."[14] Meany's rhetoric aside, public employees—led by AFSCME's Jerry Wurf—were at that moment (the Watergate scandal had given Democrats enormous majorities in Congress) most optimistic about extending coverage of the Wagner Act to public employees.[15] The effort attracted vicious assaults from right-wing ideologues, however, and its proponents lost all hope of a new law when the Supreme Court struck down federal jurisdiction over state and local labor relations in 1976.[16] Still, given all of the militancy in the public sector—with teachers prominently represented—labor's agenda could certainly have further heightened fears of turmoil.

The Crisis

By late 1974, New York City faced a budget deficit of epic proportions. Although the practice began during Robert Wagner Jr.'s tenure as mayor, Republican John Lindsay increasingly relied on short-term loans to pay the city's expenses, using future "revenue anticipation" as collateral. Paying forward the city's costs worked for a time, but, like much of the rest of the nation, New York City had been hard hit by the economic downturn that began in late 1973. Coupled with the city's long-term deindustrialization—according to a Congressional Budget Office report, the city lost nearly 400,000 jobs from 1970 to 1975, including 80,000 in 1974 alone—the recession caused a bigger shortfall of revenue while demand for social services increased. Elected in November 1973, Democrat Abraham Beame revealed less than a year later that the city's budget was more than $300 million short and announced the layoffs of 3,700 city employees and the mandatory retirement of another 2,700.[17]

Unemployment in the city ballooned to double digits the next January and to almost 12 percent—its highest level since the Great Depression—by June. The poor economy eroded tax revenues, and banks grew anxious about further extending credit. By the end of February 1975, New York City's total debt—$7.9 billion in long-term obligations and $5.7 billion short term—was larger than the annual city budget, which itself was almost $900 million more than the city's anticipated revenue for the 1975–76 fiscal year. The city struggled to sell short-term notes at increasingly higher interest rates in late 1974 and 1975, and by April, after Standard and Poor's downgraded the city's credit, the bond market was effectively closed in New York.[18]

It should be noted that even as Standard and Poor's downgrade signified banks' new confrontational stance, Moody's—the other major credit agency—upheld the city's rating, noting that "for a half a century now, it has been widely known that New York City has ... a systemic difficulty in raising additional revenues to keep up with expanded needs" but that "New York City's debt is secured by much more than its current liquidity position."[19] Indeed, while the aggregate amount of debt was indeed large, the very size of New York City's economy—as large as many world nations—might have afforded a much less austere response.

The banks' position, however, meant the city needed infusions of money, and the Ford administration was not interested in providing it. In May, Ford responded to Beame's entreaties that "a federal [loan] guarantee by itself would provide no real solution, but merely postpone for that period [of the guarantee], rather than coming to grips with the problem."[20] The State of New York—its own credit rating tied directly to the city's—had a much more compelling interest in preventing default and expedited a $400 million direct-aid payment. Short of federal help, however, the city still needed to convince investors to buy more bonds. Thus, Beame introduced a series of steep cuts to "put the city on a sound fiscal footing." Even after laying off over 500 police officers, 300 firefighters, and almost 800 sanitation workers, as well as closing four municipal hospitals and 43 schools, however, New York City still faced a deficit of almost $650 million in late April.[21]

After upstate Republicans in the legislature refused to authorize the city to raise taxes, Beame gave an emotional, televised speech—he reportedly wept—at the city council chamber where he outlined an "austerity" budget, which by the end of July would cut 5,000 teachers; another 4,000 cops; 5,000 municipal hospital workers; 1,000 more sanitation workers; and 565 additional firefighters.[22] By the end of May, it was clear that even with the cuts, the city would still default without outside assistance. At this point, Democratic governor Hugh Carey brokered—with the city's largest banks—the formation of a new state agency called the Municipal Assistance Corporation (MAC; colloquially referred to as "Big Mac"). Consisting of five members with ties to finance or banking, including the chairman—Felix Rohatyn—of the powerful investment banking firm Lazard Frères, Big Mac would help the city make loan payments in exchange for significant leverage over the city's fiscal policy.

Meanwhile, the Patrolmen's Benevolent Association (PBA) and the Uniformed Firemen's Association (UFA) attempted a public-relations campaign to pressure Beame and MAC from carrying through on cuts to the police and fire departments. In tourist entrepôts like Kennedy Airport and Penn Station, police officers and firefighters passed out skull-decorated pamphlets titled "Welcome to Fear City" and leaflets with the headline "If You Haven't Been Mugged Yet . . ." that cited crime statistics in Gotham." In the context of heightened fears of crime exacerbated by the economic downturn (in 1974, according to a U.S. Department of Justice study, "serious crimes" in New York City increased by 12 percent over the previous year, while, nationwide, criminal activities increased by their largest rate since 1960), the unions hoped to convince New Yorkers of the cuts' disastrousness.[23]

To protest pending layoffs, sanitation workers at the end of June in Manhattan and Brooklyn staged wildcat strikes leading to garbage pileups. This strike happened at the exact same time that 76,000 AFSCME workers in Pennsylvania began a four-day strike against a governor who had run on lowering taxes, and Seattle voters went to the polls to decide whether to recall Democratic mayor Wes Ullman for laying off firefighters in response to a fiscal crisis there.[24] The New York wildcat strike spread quickly, and for almost a week, the media showcased images of uncollected garbage. Several hundred newly laid-off police officers shut down traffic on the Brooklyn Bridge in protest on July 1, and the next day, firefighters called in sick in "unusually high" numbers. The *New York Post* on July 2 concisely summed up the developments with the headline "IT GETS WORSE." The garbage strike ended with Albany allowing the city to raise a new series of one-year taxes in order to rehire half of the sanitation workers scheduled for layoffs, but disgruntled citizens continued to set piles of garbage aflame before it could be collected, and the informal sick-out limited the firefighters' available to extinguish them. The negotiations between Beame and the state also allowed the city to rehire a large portion of the laid-off police officers and firefighters, but the capitulation met with MAC's disapproval.[25]

After bonds nose-dived in late July, MAC called for more austerity—a new round of permanent layoffs, an increase in the subway fare, and the end of free tuition at city universities—in order to reinstill faith in the city's credit.[26] These calls clearly signified an intent to reshape the city. In the words of one *New York Post* reporter who commented on the higher-education cuts, "free tuition has as much to do with symbolism as with dollars. To the city it is a generous 128-year tradition of educational opportunity and represents a helping hand to generations of upwardly mobile citizens. To many suburban and upstate residents it is an escalating tax burden that they should not be forced to share."[27] The cuts to the City University of New York (CUNY), in particular, revolved around race, since the free university system disproportionately offered opportunities to black and Puerto Rican New Yorkers. Indeed, according to a young Al Sharpton—then chairman of the National Youth Movement—74 percent of the students who received free tuition in the CUNY system were black or Puerto Rican.[28]

On July 31, Beame further capitulated to MAC's demands in another highly emotional speech asking the city council to empower him to institute a total wage freeze for city employees—"if not voluntary, then imposed"— even in contracts already negotiated. After meeting with the Metropolitan Transit Authority (MTA), Beame also announced that subway fares would be increased from thirty-five cents to fifty cents, while CUNY's budget would be pared $32 million. By August, Governor Carey and MAC almost completely controlled the city's ability to maneuver, and in early September, Beame officially ceded control of New York City's finances to a seven-person Emergency Financial Control Board (EFCB), appointed by Carey, that would work with city and state pension funds to save the city from default.[29]

As a result of the education cuts, teachers went on strike in September. Although UFT president Shanker (now president of the AFT as well, defeating Selden in a 1974 election) privately opposed the strike, he responded to the overwhelming calls from the rank and file to strike to restore jobs to laid-off teachers. Making lemonade out of lemons, Shanker negotiated the rehiring of several thousand teachers by using the fines the union would be forced to pay for violating the Taylor law. The EFCB opposed this settlement, however, and as a result, Shanker used the only leverage left at his disposal: refusing a $150 million contribution from the Teachers Retirement System (TRS) in a standoff that brought the city within an hour of defaulting in October.

Finally, after a lengthy public-relations campaign by Beame and Carey— who toured the country discussing the national consequences of default—

and a public warrant to act (a Harris poll that November had 69 percent support, including 62 percent of Republicans, for a loan guarantee), Ford was willing to back New York City paper. Still, the president agreed to the package only after the state legislature forced Beame to lay off another 8,000 city employees in early November, the state and the city worked together to pass a $1 billion tax package, and the city's pension funds agreed to further investments in New York debt, including an additional $1 billion from TRS. When signed into law, the New York City Seasonal Financing Act of 1975 provided the city a series of short-term loans—$1.3 billion in 1976 and $2.1 billion in both 1977 and 1978—under the condition that it balance its budget by 1977.[30]

By the end of 1975, the immediate threat of default had been vanquished, but the city's economic and social order had been set on a very different trajectory. By the end of September 1975, the city's workforce had been reduced by 30,000 through layoffs, and would be reduced by another 60,000 through attrition by 1980. Not only were almost 100,000 good-paying jobs with benefits lost for New Yorkers, but those who remained would also need to work more while their salaries were frozen during a time of rampant inflation. From late 1974 until the end of 1976 alone, the city lost 25 percent of its teachers, and class sizes reached fifty in some places, prompting student walkouts in the 1975–76 school year. The cuts disproportionately hurt the city's African American and Puerto Rican school districts. CUNY discontinued its policy of free tuition, enrollment declined by more than 60,000 students from 1975 to 1980, and faculty was reduced by 3,000. Ironically, the pension funds of city employees provided the real "bailout" to the city: by the end of 1977, these funds held $2 billion (or just over 20 percent) of the city's debt, while the federal government held only 13 percent.[31]

Mayor Beame Points Out the City's Special Burden

Mostly disregarded in accounts of the fiscal crisis as a visionless defender of the status quo, Mayor Abraham Beame deserves more credit for his effort to save the robust social-welfare state and public sector in the city. For much of 1975, in fact, the mayor argued vehemently for a grander structural solution to the city's financial woes. In this effort, Beame's view was similar to prominent liberals like the Keynesian economist John Kenneth Galbraith, who argued that "it's fair to say that no problem associated with New York City could not be solved by providing more money. The remarkable thing is not that this city's government costs so much, but that so many people of wealth have left. It's outrageous that the development of the metropolitan community has been organized with escape hatches that allow people to enjoy the proximity of the city while not paying their share of taxes."[32]

In early 1975, the mayor highlighted several factors responsible for the city's financial problems. Like Galbraith, he pointed to the particular expenditures necessary in administering a city that also supported the suburbs, and he also showed that, because of how the State of New York structured its welfare payments, the city paid out more in benefits per capita than most other cities in the United States. Second, Beame emphasized the effects of long-term deindustrialization and the immediate economic downturn that sliced revenues while costing the city more in social services. Critics of Beame could certainly point to his political motives in highlighting these developments instead of poor city management, of which he had played a part as Lindsay's comptroller from 1969 to 1973. Furthermore, trying to score aid from either the state or the federal government clearly represented a more palatable solution to the fiscal crisis than cutting services and laying off workers. These points cannot be refuted; however, whatever fiscal shenanigans took place do not preclude an explanation for Gotham's problems that includes a critique of the structure of the metropolitan area. In other words, city leaders could have covered up poor accounting practices for political reasons at the same time that structural problems caused the financial shortfall in the first place. Beame clearly thought that pointing out the city's overarching limitations bore a good chance of working—largely because independent reviews of the fiscal crisis substantiated this explanation. Indeed, a Congressional Research Service account of the events in October 1975 legitimated most of Beame's critiques.[33]

In any event, Beame's rhetoric serves as an important bellwether. Only after the possibility of larger interventions became impossible—as MAC's calls for concessions from the city's workforce and social services represented the only way the city could avoid default in the spring and summer of 1975—did Beame join the chorus of attacks on the city's workforce for causing New York's fiscal problems.

In January 1975, Beame testified before the Senate Subcommittee on Inter-governmental Relations. In front of a Democrat-majority committee, he called for a total federal takeover of payments for both Aid to Families with Dependent Children (AFDC) and Medicaid in addition to increased revenue sharing for New York City on the grounds that the current formula was "discriminatory." Criticizing recent cutbacks in urban spending, Beame asserted that these policies had "simply transfer[red] federal responsibilities to local governments and impose[d] additional burdens on local taxpayers." The city's high unemployment rate resulted in part from Ford's tighter monetary policy, he argued, and also helped to usher in the city's financial difficulties. The spike in unemployment cost New York City $150 million in tax collections during the second half of 1974, according to Beame, in addition to "an unexpected increase in welfare and Medicaid costs caused by the recession."[34]

In calling for federal assumption of all costs, Beame argued that "welfare is not a local responsibility any longer." Rather, the federal-state-local sharing of costs represented an outdated vestige of a time when people were less mobile: "The same problems which beset New York City, because of large migrations of people from agricultural areas, have also become the problem of most central cities in the nation. The cities were not responsible for the interstate movement of people in search of employment."[35]

In addition, Beame argued that the city did not receive its rightful portion of federal revenue sharing under the State and Local Fiscal Assistance Act of 1972. Signed by President Nixon, the law was designed to offset the difficulties faced by cities in the postwar United States. As explained in a policy report by the Staff of the Joint Committee on Internal Revenue Taxation (a bipartisan, bicameral committee), the new law responded to several developments:

> Population growth generally, and urbanization especially, have increased many fold the need for more extensive services. This increased need is clearly evidenced in the case of services such as police and fire protection, refuse disposal, sewage systems and street and mass transit systems—expenditure categories which tend to increase most rapidly with urbanization. The cost of these and related categories have risen from $7.7 billion in 1957 to $17.8 billion in 1970. The inflation which has been experienced in recent years has added greatly to costs.[36]

The problem with the law, to Beame, was its "ceiling": a city could only receive 145 percent of the average aid for municipalities and counties in that state. This ultimate limit kept New York—arguably the city with the greatest need in the nation—from almost $300 million annually in funds that the complicated formula (based on population, poverty, and level of urbanization) would have otherwise allotted it.[37]

In early March, Beame pressed the city's case in U.S. Senate Budget Committee hearings. He asked the committee to consider a more substantial federal-aid package for New York and other cities to set right the "great inequities [that] have developed in the fiscal interrelationships among federal, state, and local levels of government in the United States." Referencing Ford's recent proposal to cut income taxes (while also cutting spending to limit inflation), Beame argued that there needed to be a "coordinated, intergovernmental approach" to local problems instead of the current state of affairs in which "local governments are being forced to raise taxes, while the federal budget proposes to stimulate the private sector through tax deductions . . . and local governments are being forced to cut public service in health and welfare." As evidence, Beame showed how the costs of state and local governments had exploded since World War II as a result of increased spending on education and urban infrastructure as well as contributions to federal programs like Medicaid and AFDC. From 1946 to 1974, he pointed out, state-local expenditures increased from just over 5 percent of GDP to almost 15 percent, while federal expenditures rose only from 16.5 to 18 percent, and much of that increase was baked into the military budget. Beame thus argued for an expansion of the Comprehensive Employment and Training Act's public-employment programs by $5 billion, a $3 billion appropriation for a federal public-works program, and a new one-year $5 billion emergency-recovery measure for local governments, to be renewed in 1976 if the economy did not rebound. "I realize," he concluded, "that the above proposals add up to billions of dollars. But, Senators, the cities of this country have major problems delivering government services which the public needs and wants, and we cannot do it on our tax base alone."[38] Indeed, perhaps most significant about his early efforts to deal with the crisis is that in spite of cutting city payroll, Beame did not blame what others would call extortive unions, excessive pensions, or social welfare.

Criticism of Public-Sector Unions

As we have seen, unions in New York City pioneered public-sector collective bargaining. Their efforts often required strikes, and teachers were involved in some of the most controversial. By the time public employees aggressively defended their jobs from austerity measures in 1975, then, New Yorkers had witnessed many public-sector workers dramatically improve salaries and working conditions. Therefore, it is no surprise that many blamed the city's unionized workforce—especially teachers—for the fiscal crisis. On April 7, for example, Mrs. S. Schauder from Queens was "appalled" by UFT president Shanker's opposition to education cuts: "He seems to think taxpayers have to stand for ever-increasing teacher salaries and fringe benefits which the average worker can't hope to attain."[39] Brooklyn's Gloria Klinga also believed that unionized teachers had caused the city's financial problems. New York City, she argued, awarded "very lucrative contracts . . . to the teacher's union, as well as the other municipal unions. . . . Why should the taxpayer have to be burdened with their dental care, their prescription costs, and the many other extras they receive for working for this city?"[40] On April 28, O. A. Westerhaus, from Queens, wrote Beame to cheer him for cutting city services, "which are badly in need of trimming." Further, she/he pointed out that "it is a pleasure to realize that the city finally has a Mayor with the courage and fortitude to pound some sense into the heads of Shanker, Gottbaum [sic], Delury et al., before their unreasonable demands drive all the sheep that pay the taxes out of the city."[41]

The *Daily News* and the *Times*, New York's two most prominent newspapers, agreed with these assessments. The former, with a daily circulation of more than two million, was not only the largest newspaper in the city but also the most widely distributed in the United States.[42] Begun in 1919 as a tabloid "picture newspaper" designed to appeal to the masses, the *Daily News* exploded onto the scene by marketing on the New York City subway. By the mid-1920s, it became the highest-circulating newspaper in the United States, and by World War II, its daily circulation surpassed the two-million mark. The newspaper, in its appeal to the "common man," catered to the white ethnic commuters who had begun to move into the outer boroughs and suburbs, but it faced increasing competition from television after the war, and even with its sizable circulation figures, encountered financial challenges. Like most other newspapers, it sought to save money by cutting labor costs and instituting automation in the 1960s and '70s. The Newspaper Guild strike of 1962–63, which virtually blacked out print news in the city for several months, caused multiple newspapers to go out of business and raised tempers among the remaining publishers like the *Daily News*. As suburbanization shrank the pool of public-transit commuters, the newspaper tried to cut costs to the bone, and collectively bargained contracts represented an enormous impediment. Galvanized in particular by the contentious bargaining sessions in the years that followed the nine-month Newspaper Guild strike, the *Daily News*'s editorial page generally came down harshly on strikes by all workers.[43]

The *Daily News*, then, largely blamed the city's unionized public-sector workers for the fiscal crisis. On April 16, for example, the newspaper argued that "the plain truth is that New York has been on a reckless binge for a decade. It has undertaken to provide services it cannot afford: its payroll—in size and cost per employee—has been allowed to balloon outrageously; waste and inefficiency have been tolerated for political reasons . . . half measures won't help. Nothing will, except drastic retrenchment and the most ruthless pruning of expenditures."[44] On April 20, the *Daily News* asked its readers, "Think We Got Troubles?" Highlighting the negative example of Great Britain, the piece contended that excessive public-sector employment led to catastrophe.

Pointing to the United Kingdom's "belt-tightening austerity"—the Labour government had just called for significant tax increases—the *Daily News* believed that Great Britain's plight resulted from Labour's promise of "the world with a fence around it" that had "showered" workers "with benefits that Britain could not afford, and sapped their get-up-and-go with incentive-deadening spread-the-wealth schemes." Quoting the Tory Margaret Thatcher's characterization that Labour's policy represented "a typical Socialist budget with equal shares of misery for all," the newspaper concluded that "that's just what once-great Britain has come to under socialism."[45] In the context of the city's problems, the implication that public-sector unions needed to be reined in in New York was obvious.

An editorial the following week referenced the structure of metropolitan political economy in New York City and its suburban environs, acknowledging "the increasing jobless rolls, anguished cries from the cities for more aid, [and] a general acknowledgment that the property tax cannot be relied on for additional revenues in either city or suburbs." While the premises of this argument seem similar to Beame's calls for bigger structural solutions, the conclusion was startlingly different: an unconditional surrender to austerity. "A graver danger now looms," the piece concluded, "unless New York, New Jersey, and Connecticut make more than cosmetic cuts in their budgets and really hold the line on taxes. It is that firms will move out of the metropolitan region altogether and business will become an endangered species here."[46] It is not a stretch to imagine that the newspaper's editorial board projected fears about their own labor costs onto its treatment of the fiscal crisis.

The *Times*'s assessment of the crisis was basically the same as that of the *Daily News*, only with more eloquent diction. The *Times*, privately owned by the Ochs family and the nation's most prestigious paper, often put journalistic integrity ahead of profits. The newspaper, for example, had courageously published the Pentagon Papers—Daniel Ellsberg's leak of classified documents proving the American government misled the public about Vietnam—in 1971 in spite of the Nixon administration's efforts to suppress it. With a circulation of over 800,000, the *Times* was the nation's fourth-largest newspaper in 1975. Still, although it could withstand lean years, the newspaper's revenues decreased drastically in the late 1960s. According to Susan Tifft and Alex Jones, *Times* publisher Arthur "Punch" Sulzberger became a "laughingstock" among other publishers and even among labor leaders, who privately considered him 'soft'" after signing a contract in 1970 that gave workers a major salary increase without any concessions on automating the paper's publishing. After that, Sulzberger resolved not to "capitulate" again, and the newspaper's humiliation in its own dealings almost certainly affected its view of labor. In addition, the *Times* lost almost half of its advertising during the fiscal crisis as the local economy slowed down, giving its editorial board extra motivation to see the city remain solvent.[47]

Like its higher-circulating counterpart, the *Times* also called on Beame to follow the "rigorous austerity message the Labour Government has just delivered to Britons. . . . As in Britain, the fiscal crisis here demands stern measures that will be distasteful to everyone, and most of all to the municipal labor unions that have been an important source of support for the present administration."[48] By the end of May, the *Times* offered starker advice: "Though painful adjustments are inescapable, many jobs and services could be saved if Mr. Beame would face up to the city unions." Denying Beame's arguments about metropolitan political economy, the newspaper opined that "the city's fiscal crisis is not the figment of some banker's imagination. It was not made in Washington or in Albany. Implying that the real fault lies elsewhere is not the way to elicit the cooperation needed from higher levels of government or the sacrifices which will be required of all New Yorkers."[49]

Defenders of Labor Liberalism

In addition to criticism of public-sector unions, there also existed substantial discussion in the city's public sphere defending both organized labor and cherished programs. The liberal *New York Post*, which mostly agreed with Beame's structural critique of the crisis, led the charge. Founded by Alexander Hamilton in 1801, the *Post* was the longest continuously published newspaper in the United States. In the words of one career editor, it was for much of its history a "dull conservative broadsheet" until banking heiress Dorothy Schiff purchased the paper in 1939. Schiff enthusiastically supported the New Deal and turned the *Post* "into a crusading left-wing tabloid." She hired "liberal icon" James Wechsler as the paper's editor in chief, and the newspaper also featured liberal luminaries Max Lerner and labor columnist Murray Kempton. The Post supported labor through the 1960s and '70s, breaking with the other New York newspapers during the 1962–63 strike to settle with the union. Although its readership had declined somewhat—leading Schiff to sell out to Rupert Murdoch in 1976—the *Post*, at over 620,000, was still the ninth-most circulated paper in the United States, and the number-two afternoon daily, behind the *Detroit News*.[50]

In mid-April, the *Post* pointed not only to a recent surge in welfare rolls in New York City, but also highlighted, in a liberal Keynesian assumption about economic prosperity, that inflation eroded the "already enfeebled purchasing power" of those depending on the "wretched 'refuge' of welfare assistance." It called for "increased aid grants and stipends and appreciably more public employment projects" from the federal government. "Despite Washington's proclamation that the 'urban crisis' is over," the piece concluded, "its realities remain."[51] The next month, after treasury secretary William Simon declared that aid for New York City would be "inappropriate," the *Post* asked, "What happened to the concept of interdependence? Can anyone seriously argue that New York's economic ordeal is unrelated to the national economic condition? Can the prospect of national recovery be divorced from the fate of this vast state and city?"[52] Several days later, it also supported Beame's call for a federal takeover of welfare, arguing that "New York's fiscal burdens, growing heavier as it sustains a growing welfare caseload that long ago should have been shouldered by Washington, are not unique."[53]

Other voices argued for a structural critique of the fiscal crisis. In a lengthy May piece in the *New York Times*, labor editor A. H. Raskin spelled out the historical roots of the city's fiscal crisis—beyond simply blaming too much spending on city employees. Raskin sympathized with unions, writing a biography of Amalgamated Clothing Workers of America (ACWA) president Sidney Hillman in 1947 and cowriting International Ladies' Garment Workers' Union (ILGWU) president David Dubinsky's in 1977. Raskin argued:

> After World War II, welfare became institutionalized here as the technological revolution in agriculture caused Southern blacks to leave the farm for New York, attracted in part by its relatively high relief payments and its lack of residency requirements. A similar migration came from Puerto Rico. In two decades, a gigantic population exchange introduced 2 million newcomers, mostly poor and unskilled, while 2 million New Yorkers, most rich, middle-class and white, moved out to the suburbs and beyond. In this period, the city pioneered in collective bargaining for public employees, building up a host of costly contractual commitments. The revolution of rising expectations that accompanied the civil rights movement in the 1960's and the emergence of political pressure groups, both radical and conservative, added to the demand for more services, more guarantees, more money from City Hall.

Pointing out that the "biggest single outlay by far" in New York was welfare, which provided for one million people (of whom two-thirds were children), Raskin compared Gotham to Chicago: New York City yearly contributed over $1 billion for welfare, while for the Windy City, "virtually the whole bill is state and federal." Further, the city's increased workforce resulted not from union payoffs but from the growing cost of social services. Connecting mothers receiving welfare benefits with female professions like teachers and social workers, Raskin pointed out that those working in the health, education, and welfare sectors increased by almost 60,000 from 1964 to 1974 while the amount of police, fire, and sanitation-service workers had actually declined during the same period.[54]

Irate New Yorkers wrote both newspapers and public officials to defend programs that had improved their lives. In April 1975, the mayor's office received over 200 letters regarding the budget for city libraries, a number that exploded to almost 1,700 letters in May. In the last two weeks of May, Beame received almost 800 regarding hospital closings alone. From June 3 to June 15, New Yorkers sent almost 1,100 letters about the city's education budget. As the crises peaked, so did the angry letters. The mayor's office received 1,309 letters about the budget in July, and 1,700 in October. On November 3 and 4 alone, Beame received 1,481 letters.[55]

Some commentators pointed to the city's perceived level of high crime to defend cuts to the police force. Ann Volpe, for instance, writing from the Bronx, identified herself as "an over-burdened and irate tax payer in this, our 'fun city.'" In a call for law and order similar to other criticisms linking African American children and unionized teachers, she "strongly object[ed] to your relieving 400 policemen from duty. In this crime infested city such a move to reduce protection for its citizens in order to accommodate your budget is not only unwarranted but dangerous."[56]

Emil Ference defended the public library system: "It is a sad commentary on the times when our libraries are being shut down one by one. All our great men have emphasized that books and the access to books is essential to the mental and spiritual health of our democracy. . . . Save, if you must, in some other direction, but keep the libraries open."[57] The Aguilar Library Support Committee, from East Harlem, argued against cuts to the local library branch because they would hurt "the bilingual institutes, college readiness program, and a branch of the CUNY in the neighborhood whose students need and use this branch."[58]

Others defended the city's higher-education system. Free tuition in the CUNY system, argued the president of Bronx Community College in a letter to the *Times*, "is based on the principle that a college education is a citizen's right and not a privilege. . . . Only educated communities and countries are solvent. Education must, therefore, have the highest priority, especially in times of crisis."[59] Agapito Otero, a student in the CUNY system, urged Beame not to cut remedial services for city universities. He believed the programs represented "a necessity to many minorities of the City of New

York. The cutting back of these special skills programs will make a lot of students drop out, who ordinarily would not."[60]

A wave of public outcry against K–12 education cuts hit in June. Beatrice Larkin, from Brooklyn, tied the education budget to the city's demographics. "Our children are in desperate need of additional help—reading scores are down, truancy is up—yet plans are to subtract services and personnel rather than add. Whatever middle income families still remain in the City will surely leave, rather than see their children in overcrowded classrooms with fewer teachers and little or no services."[61] Dorothy Gunderson, from Staten Island, wrote Beame to remind him of the unfair distribution of tax revenues in New York: "Let me state that New York City taxpayers contribute most of the money to the State treasury and in return receive a disproportionate share of the tax money collected. . . . I ask that you do everything in your power to obtain more state aid in order that the crisis budget for New York City will not be executed and our school system saved."[62]

Welfare Chiselers and the Decline of the Middle Class

Many New Yorkers who considered themselves "middle class" blamed welfare recipients—whom they believed received illegitimate benefits—for the city's fiscal crisis. Further, this critique centered on a racialized and gendered view of welfare and a characterization of New York as degenerating into chaos because those who produced had to support the "unproductive." Although some of these critics believed such excesses came at the expense of the city's unionized workforce, far more, particularly as the year went on, explicitly connected city workers, particularly teachers, and welfare recipients. This view existed before the July labor actions of the city's blue-collar unions. Those events, however, coupled with the UFT strike in September, lent focus to the recalculation of political allegiance among the city's productive middle-class "backbone." Indeed, in categorizing unionized public employees outside the boundaries of "middle class," this amorphous constituency reshaped the contours of how the city's social order was understood.

In February, a "cooperative apartment owner strained to the limit" in Manhattan expressed "dismay" following Beame's call for higher real estate taxes. "Again," R. W. Houseman believed, "the middle income tax payer will suffer for the wealthy and the poor. How much can we take in this city! It is already impossible to cope with the physical problems of living in New York City—expense, crime, dirt, lack of any decent civic services etc."[63] Criticizing potential cuts to garbage collection, Bill Friel defended the homeowner, calling the single-family home the "backbone of the city. We are the ones who pay our taxes, donate our time to various causes, sweep our sidewalks and the gutter in front of our homes. From daily pickup to only weekly over the years is just asking homeowners to rebel."[64]

Other "middle-class" citizens called explicitly to stanch welfare spending. Although welfare could have signaled any number of social services, by the early 1970s, it referred almost exclusively to the AFDC program. Because of the stigmas associated with "welfare" and discriminatory administration of the program, in the early 1960s, only about one-third of families eligible for benefits received them. In the mid-1960s, however, Congress loosened the rules for eligibility, the number of those on welfare increased, and the National Welfare Rights Organization (NWRO) organized for better benefits and an end to the humiliating assumptions about mothers on welfare. Indeed, the NWRO consciously lodged an effort to destigmatize a program seen by many whites as one in which black women's behavior needed to be constantly surveilled and controlled.[65] Some welfare rights activists even believed that the increase in the welfare rolls could precipitate a crisis within the Democratic Party that would force a systemic response to inner-city poverty. As Richard Cloward and Frances Fox Piven argued in 1966, a dramatic increase of the welfare rolls would "deepen existing divisions among elements in the big-city Democratic coalitions: the remaining white middle class, the white working-class ethnic groups and the growing minority poor. To avoid a further weakening of that historic coalition, a national Democratic administration would be constrained to advance a federal solution to poverty that would override local welfare failures, local class and racial conflicts and local revenue dilemmas."[66]

Although the Democratic Party spectacularly failed to "avoid a further weakening" of the New Deal coalition, those receiving AFDC benefits expanded dramatically in a short period of time. In 1966–67 alone, the number of recipients increased by 600,000, and, by 1970, there were almost 12.5 million Americans on welfare; 25 percent of the increase occurred in California and New York. Jill Quadagno points out, "The costs of skyrocketing welfare rolls were primarily borne by working- and middle-class taxpayers, whose tempers rose in tempo with welfare costs. Why, they wondered, should families struggling to make ends meet pay for loafers who lived off the government's largesse?"[67] Linking the new visibility of the African American poor with fears of urban insurrection, many middle-class whites vehemently reacted against welfare; whereas 60 percent of Americans supported increased welfare spending in 1961, that number had fallen to 20 percent in 1973.[68]

In this context, many of those "middle-class" New Yorkers tried to explain the crisis by pointing to their victimization at the hands of welfare recipients.

One Brooklynite, for example, asserted in a letter to the *Post* published on April 1 that "all Mayor Beame and Gov. Carey know how to do is raise taxes and load them on the middle-class working people. Why don't they stop all those giveaway programs and go after the welfare chiselers?"[69] It enraged Helena Hauerstein, from Manhattan, to witness the mayor "begging for money." Instead, she wanted him to "crack . . . down on incompetency and chiselers in the welfare set-up."[70]

One critic articulated an apocalyptic narrative to explain how excessive welfare had shaken the city's social order. In a letter that could have been written by Travis Bickle, the paranoid protagonist of Martin Scorsese's earth-shattering film *Taxi Driver* (1976), Marion Wertheimer complained that the June subway fare increase forced her "to move from my old neighbor-hood because it became a jungle—noise—filth—theft. I am alone now—not married—had to pay double my rent in order to at least survive and now am in a two fare zone to boot." If the racial coding of "jungle" and "filth" wasn't already apparent, she explained to Beame that "the big bottle neck is all the welfare being paid out for people who do not want to work. . . . I have heard of many cases where the city pays triple the rent for apartments when they move in a welfare participant."[71]

Others explained the crisis by specifically linking welfare recipients and public-sector workers into a unitary group of nonproductive freeloaders. Jacob Beller, for instance, believed that "what is happening in New York City now is all the good. Welfare fraud and waste; educational fraud and waste; no-show jobs; 'hungry' unions; 'crazy' pensions; all have to come to an end."[72] Similarly, Brooklynite Maureen Cullen, a "resident and a tax-payer of the City of New York," believed that both welfare recipients and public-sector workers shouldered the blame for the city's new slogan of "BANKRUPT CITY, [WHERE] CRIME AND FILTH ARE ON THE RISE." She believed Beame played "a dirty game at the expense of those who work hard for a living and pay our taxes." Welfare represented "one of the most abused areas. I personally resent seeing my money spent on welfare, and I can as-sure you that most working people feel the same way." Cullen proposed specific changes to reform it. First, "you should be a resident of the city for at least five years, no exceptions made, to even qualify." Second, "a complete personal investigation should be made to determine if the person is truly justified in applying for welfare," and, finally, anticipating the trajectory of AFDC following the passage of the Personal Responsibility and Work Op-portunity Reconciliation Act in 1996, she argued that "a limited time for collecting welfare benefits should be set and upon completion of this time period, checks, food stamps, etc. should automatically be discontinued."

Teachers, in her estimation, were also takers: Cullen wondered, "Why are teachers paid for a two month vacation during the summer months? They should be given a regular three week vacation, and then be assigned to various departments within the City, thereby eliminating overtime."[73] Indeed, Cullen clearly assumed that unproductive welfare recipients and teachers alike appropriated the hard-won earnings of taxpayers like herself.

Other voices, particularly in the crisis's early months, however, defended welfare. From Brooklyn, Sherian Ellis believed "it is unfair to assume that people on welfare are good-for-nothings. If given a break, a lot of the less fortunate would gladly drop off the relief rolls. It is too easy for many of us to point the finger of accusation at those on welfare."[74] *New York Post* columnist Harriet Van Horne illuminated the larger economic problems the city's poor faced in 1975. Beame's April and May cuts only made city life more difficult, she argued, causing a "quickening" decay and a "rot" that was "moving into middle-class areas"; she also compared the United States to its Anglophone friends across the Atlantic: "Some economists see Great Britain as the prototype of eventual U.S. decline. We are entering a stagnant period in which people who are working will pay heavier taxes to support people who are not. But we are beginning to see these people more clearly. They are not the 'lazy welfare cheats' who loom so large in Republican rhetoric. They are skilled artisans, white-collar management, teachers, and policemen. June graduates and bright teenagers idle for the summer. They are all scared—and their patience is running out."[75]

Earl Graves, writing for the *New York Amsterdam News*, underscored the "black stake" in the crisis. He argued that "the typical Black New Yorker has much less of a realistic option of leaving the city, no matter how bad things get, than his typical white fellow citizen. He is much more dependent on the everyday services provided by the city, such as public transportation, garbage collection, and police and fire protection." Graves also pointed out that blacks made up about 75,000 of the city's 300,000 workers and were "disproportionately numerous among the city's lower paid workers, so that they form a particularly vulnerable group within a vulnerable group to begin with." Finally, "as one moves down the economic scale, the figures for the Black share become staggeringly disproportionate. . . . Of the nearly one million New Yorkers who receive public assistance, to which the city contributes some $650 million a year, 43 per cent are Black."[76]

Figures as diverse as Vernon Jordan, executive director of the National Urban League, and Albert Shanker argued for the federalization of welfare to alleviate the fiscal crisis. Jordan argued that New York "pays over a billion dollars on welfare costs that should be federalized because poverty is a national problem caused by the failure of the economy to generate jobs for all. . . . The solution to city financing problems won't be found in closing schools and firehouses, it will be found in determined national policies to create jobs and get the economy operating in high gear again."[77] Shanker also blamed the federal government. Like Jordan, he explained the large increase in the city's welfare rolls as the result of economic recession and the failure of the Nixon and Ford administrations to get the economy going again.[78]

Criticism of welfare recipients, then, was heavily contested. Prominent voices continued to point out the structural reasons for the city's gigantic spending on programs like AFDC. Still, the discursive efforts to link teachers and welfare recipients as equally culpable for wasting "middle-class" tax dollars had begun to gain steam before the July blue-collar actions and the UFT strike in September.

Unions Behaving Badly

By early July, a raging debate about who caused the crisis and how it should be solved had burned for months. Although many criticized public-sector workers, the structural explanation for the city's problems was still alive. And though many criticized the illegitimacy of welfare recipients and linked them to public-sector workers, it was far from inevitable that a narrative in which too many unproductive New Yorkers siphoned away the city's revenues would emerge as the most prominent. Ironically, the sanitation strike, police protests, and firefighter slowdown in early July combined with the teacher strike in September and near default in October to allow this explanation to drown out all others. These unions—the blue-collar unions defending the sanctity of signed contracts and the teachers trying to avert stark cuts to education spending—turned to drastic actions that did stave off even worse scenarios for their members. These desperate acts, however, allowed the unions' opponents to argue more fervently that its employees did not care about the city's future.

Only after exhausting all other avenues did Beame seek concessions from the city's unions. The banks forced the mayor—if he wanted to keep the city solvent—to try to force unions to accept drastic cuts. As a result, Beame's rhetoric shifted; in his July 31 call for alterations to contracts, Beame's characterization of the city's workforce had changed. The mayor first called for a "wage freeze—if not voluntary, then imposed." More significant, Beame proclaimed that "in the future of collective bargaining, we will concentrate on what comes out of a contract as well as what goes in. . . . Excesses should be eliminated in future contracts."[79] Taken together, these two statements indicated not only that workers had to be reined in for fiscal solvency but also—an argument the mayor had not yet made—that "excesses" in union contracts had helped cause the crisis. Whether Beame believed this explanation or simply capitulated to what he thought investors wanted to hear is irrelevant. Mayor Beame—the city's most prominent defender of a structural explanation for the crisis—now placed the excesses of unions at the center of the story.

The *Daily News* led the charge against public-sector unions in the media, in early July rebuking the wildcat sanitation strike. The newspaper responded dismissively to workers' claims that they were defending their collective livelihood: "Let's get one point straight right off. The sanitmen aren't seeking merely just or fair treatment. Instead they are insisting on privileged status, and using the tactics of the bully and the blackguard to obtain it. They are deliberately creating a menace to public health to beat New York into submission." The *Daily News* admonished Beame to resist, averring that "if City Hall budges an inch or makes a single concession, it will be offering the citizens of New York as a punching bag for every greedy, arrogant union in the city."[80]

The *Post* turned against the sanitation workers, too. The newspaper called the garbage strike—dubbed the "Stink City" campaign by the Sanitationmen's Union head John Delury—a "man-made condition that could actually imperil the health of thousands of men, women, and children. . . . Delury's unforgettable phrase has polluted the atmosphere more permanently than any uncollected can of garbage. It will long be cherished and quoted by anti-union lobbyists throughout the country."[81] James Wechsler's editorial page unequivocally upbraided the police as well. Calling the "Fear City" campaign "a dangerous, irresponsible attempt to intimidate city officials," the newspaper predicted that "the experiment will backfire."[82]

Individual observers also criticized workers for striking, and, like Beame, began to shift their explanations of the crisis to focus on the high costs of municipal workers. Indeed, as had happened in the discussion over the desperate plight of Philly schools in 1973, the strikes and sick-outs seemed to draw attention to the perception that workers' contracts caused the city's near bankruptcy. Kendall Lutes, from the Upper West Side in Manhattan, wrote the mayor that "enough is enough . . . the private citizen cannot go on bearing the unreasonable cost of these municipal unions." The crisis also spurred Lutes to action, and he planned "to organize a protest to these illegal acts on the part of the unions. If we let these municipal unions continue we will never get this city out of its financial problems. Please let me know what you are going to do about this outrage."[83] John Murray, a father of five from the Queens neighborhood of Jackson Heights, wrote Beame to inform him of his political consciousness. "I have never written to an elected official of the city government," he began. "My first letter." His anxiety about the future of the city, however, compelled him to do so: "You cannot and should not bow to the pressure of the city union leaders to put back employees we cannot afford and then ask me and my children to support this union muscle for many years to come with unnecessary taxes. Please . . . we can't handle it."[84]

If criticism of city unions reached inferno level after the July conflicts, the teacher strike in September caused a veritable explosion. The strike occurred at the high-water mark of teacher militancy in the United States; indeed, like New York, other cities faced diminishing tax revenues, high inflation, and a militant public sector. During the entire 1974–75 school year, there were 99 teacher strikes in the United States; during the course of the 1975–76 school year, there were 203, and by September 22, 1975, there had already been 106.[85] In addition to the UFT strike, teachers in Chicago walked out on September 3 over salaries and class sizes; that strike—Chicago's fourth in six years—lasted eleven days. Teachers received 7 percent salary increases, and the state's Democratic governor was left duking it out with Mayor Daley's supporters in the legislature over additional state aid and tax increases.[86] There were also strikes in Pennsylvania, Rhode Island, Massachusetts, Ohio, and Montana.[87] All told, by the first week of September, teachers were out in eleven states, idling one million students.[88]

When New York City's teachers walked out the second week, two million American students were out of class.[89] Boston teachers went on strike for a week on September 23 as a consequence of Judge Arthur Garrity's controversial desegregation order.[90] The issue was that teachers had to put in extra time for meetings and counseling sessions and wanted pay raises to compensate; they won a 6 percent salary increase.[91] In Wilmington, Delaware, teachers left classrooms for twenty-eight days—the longest public-sector strike in the First State's history—and local authorities arrested more than 200 for contempt of court after teachers disobeyed the state's antistrike law. Teachers won an 18 percent raise over three years, however, and a "no-reprisal" clause that reinstated teachers fired for striking.[92] In October, a third of Atlanta's public school employees walked out for collective-bargaining rights; a judge successfully ordered them back to work.[93] Finally, in November, teachers in New Haven, Connecticut, walked out for two weeks; ninety went to jail—some for as long as a week—and the strike ended after the city's unions threatened a general strike. Teachers got modest raises.[94]

In this context—especially in Gotham—opponents asserted that unionized teachers had gone too far. Mayor Beame, for example, in remarks on
September 10, nominally argued that teachers were not directly to blame:
"The dilemma of our city really has no villains—only victims. . . . The tragic
strike now affecting the city school system is symbolic of the hard decisions
our current crisis has forced upon all of us." Still, the mayor also underscored
that teachers defied the law, arguing that they "must recognize that they are
acting in clear violation of the law which the city must and shall uphold."[95]

The *Daily News* continued its vociferous opposition. Its September 10
editorial page, for instance, featured both a cartoon and an editorial. The
former was titled "To Hell with the City," and an apelike Albert Shanker
held up a sign that read simply "WE WANT EVERYTHING! UFT" while two
white schoolchildren—a curious choice given New York City's school demographics—looked on. The editorial argued that if Shanker "really cared
about holding down class sizes at a time when the Board of Education is
struggling to make ends meet, he would willingly surrender some of the
featherbedding privileges the board is trying to end."[96]

The *Times* did not characterize the teachers sympathetically either. Even
before the strike, the newspaper equated Shanker's criticism of education
cuts with the PBA's "Fear City" tactics: "In predicting that the austerity
budget will leave classrooms too crowded for learning and the schools
abandoned to violence, the president of the United Federation of Teachers
is trying to persuade parents that their children can be saved from chaos
only by surrender to his union's excessive demands." The *Times* asserted
instead that "the sacrifices to be asked of teachers amount to little more
than the discontinuation of luxuries in staffing which should never have
been approved and which the city can no longer afford."[97] The *Post* also
followed suit. In the September 9 edition, the newspaper called the strike
"as morally unconscionable as it is technically illegal. It comes at a moment
when the city is fighting for its life and when the spectacle of a paralyzed,
chaotic school system will play into the hands of those in Washington and
other places who are depicting New York as a doomed island. . . . Rarely
has any group embarked on a strike with so little present or prospective
public sympathy."[98]

Teachers defended themselves, but the boundaries of the battle lines still
pitted teachers' interests against those of the public. Former AFT president
David Selden attacked the *Times*'s strike coverage, finding it "not only regres-
sive but completely unrelated to reality." Pointing out that he had "no reason
to speak kindly of Albert Shanker and his works"—Selden had lost the AFT
presidency in bitter fashion—he believed that "if [Shanker] has erred in these

negotiations with the Board of Education it has been on the side of being too soft, rather than too hard." The teachers' concessions "would be unthinkable in any other negotiations in public or private industry." Selden also argued that among the "New York City establishment, most of the members of which do not live in the city and whose children go to suburban and private schools, has no real interest in the school system. The teachers have introduced reality into the school crisis by the only means they have: the strike."[99]

At this point, citizens also honed their disapprobation on teachers. A. A. Legge from Manhattan, for example, wrote to the *Daily News* on September 7 that "teachers are worth about half their present wages."[100] Herbert Ewert from Queens used the language of "productivity" in defining his interests: "The taxpayer has no voice in what is going on in the city. We don't want school hours cut." Connecting teachers to social services, he argued that "these overpaid, non-productive teachers will want summer unemployment insurance and food stamps next."[101]

Furthermore, although African American activist groups worked to avert the worst of the fiscal crisis, many clearly did not view their interests in alignment with the UFT.[102] The conflict over community control in the late 1960s left much of the African American population embittered, and they did not support the UFT in 1975. The *New York Amsterdam News*, for instance, condemned the UFT's demands, arguing that instead of agreeing to concessions in order to prevent a default, "Al Shanker seems bent on using the power granted him through the loyalty of dues-paying Black peoples of the nation to blunt the law of the land, cripple New York City, and deny little Black children their rightful heritage."[103]

Conclusion

The eleventh-hour escape from default—in which Shanker dramatically restrained approval for the TRS purchase of city paper until the last possible hour—provided the last climax before the crisis's dramatic resolution in November. The *Daily News*'s version of events perhaps best shows how far the explanation of the fiscal crisis had shifted to one in which the city's unionized workforce—especially the teachers—represented the primary culprit. There was nothing that required the municipal unions to invest their members' pension funds into city paper. In fact, in a climate in which Ford explicitly and repeatedly denied federal assistance, investing $150 million in Big Mac bonds represented a serious risk. Although Shanker and the pension-fund trustees invested an enormous amount of money in the city's future, the newspaper nevertheless skewered the AFT president for "h[olding] back so long, in a cynical power play, that the city teetered on the brink of default. . . . It was an unconscionable gamble that toyed with the lives of thousands of municipal workers."[104]

Three final comments on the crisis as it lurched toward its conclusion evidence how significant the year's events had been in pushing many New Yorkers to turn against the labor liberalism that had defined the city's postwar history. On October 16, just days before the city came closest to defaulting, *ABC Evening News*'s Howard Smith attempted to defend the city's liberal tradition. New York City, he argued, in a racially charged structural explanation for the crisis,

> has carried a burden no other city on earth has carried. For a century it's been the lone place where millions upon millions of immigrants have entered, and settled, for a while. . . . New York carried the load, gave them jobs, let them save, 'til their children could move on and be admirable taxpayers. . . . The city's latest wave of immigrants is the hardest to sustain and assimilate of all—millions of Puerto Ricans and blacks who could find no jobs at home . . . and so came to New York for its unique reputation for processing poverty into means and turning its tax-eaters into other states' taxpayers.

It was clear from this critique that liberals like Smith had all but internalized the view that New York had created a "forgivable sin" in providing decent living standards to its citizens. The city, he argued, "has been over-generous to the poor and low-paid and must be restrained."[105]

The second comment shows how so many of those who defined themselves as middle class believe that the proliferation of "unproductive" citizens led to the collapse of postwar American prosperity. John Santore, from the Queens neighborhood of Richmond Hill, was a "disheartened, disgusted New Yorker." He wrote to Beame—just before the federal loan guarantee passed Congress—that the mayor and other city politicians were guilty of "ruining a once proud and great city." In language very different from that of Smith—but with much the same assumptions—he argued:

> By creating a welfare dynasty, you attracted to this city the dregs of all races. We fear for our children in school. We cannot walk the streets without fear. We cannot ride the subways without fear. Our streets are getting dirtier day by day. We now bear a heavy tax load. . . . Outrageous union contracts are inflicted upon us. . . . I never thought the day would come that I would think about leaving my hometown and my father's birthplace. Believe me, the urge gets stronger every day.[106]

Santore's explanation for the crisis speaks to the root of the delegitimation of labor liberalism in New York City. Tellingly, he invoked the concept of "we" in repeated opposition to the racialized recipients of welfare, criminals, and public-sector labor unions. By defining his origins through his father, furthermore, he made clear the gendered nature of his claim regarding the threats to the city's productive. Indeed, Santore sensed that as a white middle-class American man—a central component of the labor-liberal coalition—he was now being victimized by a liberal state allied with the excessive demands of African Americans and teachers. This characterization seems to have powerfully advanced toward the mainstream of American politics in 1975.

Indeed, on November 30, 1975, ten days after declaring his candidacy for the Republican nomination, Ronald Reagan's interpretation of the fiscal crisis was of a piece with Santore's: "There is no question that the victims in New York are the three million taxpaying citizens working in the private sector who must put up all the money that pays for everything else, who for some 25-odd years had their political leaders deceive them."[107] Maybe Abraham Beame had taken a long time to come around to the views of John Santore, but Ronald Reagan had been listening clearly.

CHAPTER 5

THE PITTSBURGH TEACHER STRIKE
OF 1975–76 AND THE CRISIS
OF THE LABOR-LIBERAL COALITION

On December 1, 1975, Pittsburgh's teachers went on strike again. Under the state's public-employee law—passed just five years before—the Pittsburgh Federation of Teachers (PFT) had the right to strike against the board of public education. In addition to allowing the formation of public-sector unions, Pennsylvania was the first of any industrial state to give public-sector workers strike rights. These rights, however, were limited; local judges could enjoin strikes if they were found to endanger the "welfare, health, and safety of the general public." Clearly, no one could easily argue a teacher strike endangered either public health or safety, but the welfare clause—as seen in Philadelphia in 1973—could be interpreted broadly. After six weeks, a local judge, on January 10, 1976, issued an injunction to force the teachers back to work in the interests of the community's "welfare."

Much of the discussion about this strike—especially before it was enjoined and the teachers violated a court order by staying out of classrooms for another two weeks—evidenced support for teachers. Many of the same supporters, however, vehemently criticized the walkout, pointing to the fundamental contradiction in the relationship between the postwar New Deal order and the public-sector labor movement: public employees—especially teachers—only partially enjoyed legal and social support for the rights private-sector workers had. The continued ineffectiveness of liberal public-sector labor law in Pittsburgh evidenced the deteriorating enthusiasm for the labor-liberal coalition in Pennsylvania.

Further, Pittsburgh represents a case study of special interest for other reasons. To begin, Pittsburgh is notable because of its long history as a central site for the American labor movement, and thus shows the emergence of a prominent anti-public-sector union discourse in a city long committed to organized labor. The Pittsburgh strike also occurred at a crucial turning point in American public-sector labor politics in the 1970s—1975 alone brought with it 478 public-employee work stoppages (prominently including those in New York, examined in chapter 4), virtually all of which were at the state or local level. More state and local public-sector workers struck in that year than any other year in American history, and teacher strikes represented some of the most spectacular conflicts.[1]

The Pittsburgh strike represented the longest and the most prominent in the wake of New York City's near bankruptcy, and it was the longest teacher strike in the 1975–76 school year. Attributed by many commentators to the purportedly insatiable demands of the city's public-sector unions, the New York crisis struck fear into smaller cities on the East Coast and in the Midwest. As I documented in chapter 4, in New York, fiscal imprudence had been both gendered and racialized through its link to teachers and welfare recipients. Pittsburgh thus represented an important test for cities coping with the new age of municipal austerity that had dawned. Even though Pittsburgh held a strong financial position, the board of public education leveraged fears about budget pressures to take a hard line against teachers. The narrative stemming from New York City made such fears more plausible. The Steel City's teachers, hoping to keep pace with inflation, emerged with substantial salary increases after a very long strike; the damage done in the city's public sphere, however, galvanized a backlash against unionized teachers.

The Context for the Strike

By the beginning of the strike in 1975, Pittsburghers, like many other Americans, were accustomed to labor conflict between teachers and school boards. As we have seen, Pittsburgh teachers went on strike in both 1968 and 1971, and the fall 1975 teacher strike wave emerged as national news many Pittsburghers read about or viewed on their televisions. Further, *Reader's Digest*—the widely circulating monthly that combined entertainment features, excerpts from pop fiction, and news stories—brought the public-sector labor problem squarely into the realm of popular culture. A piece by Irwin Ross in the October 1975 issue, titled "The Fiscal Follies of New York City," offered "a moral here for every city in the land." The piece asserted that the large deficit in New York resulted from high salaries and benefits for "its vast army of employees," about 90 percent of whom belonged to unions. The article implored "other cities [to] ponder the lessons of New York City's travail. To wit: bountiful municipal services must be limited by ability to pay; munificent pension plans make politicians popular and win union support, but the burden of financing such largess becomes enormous as the years toll on."[2] The January 1976 issue included a piece condensed from the *Wall Street Journal* titled "The Undoing of Great Britain: A Textbook Case of How to Ruin a Once-Vigorous Economy." It argued—in a cautionary tale—that generous public services and "militant trade union leaders" together were responsible for the inflation, decline of productivity, and "want of capital" plaguing the United Kingdom.[3]

The *Digest* was no mere magazine: its readership dwarfed any other periodical during the decade except the ubiquitous *TV Guide*. *Reader's Digest*'s 1976 monthly circulation of more than 18 million represented four times the circulation of the next magazine—*Time*—that featured current events. One study of the media at the time estimated that *Reader's Digest* regularly reached about 25 percent of the U.S. adult population in the 1970s. Furthermore, when one accounts for the fact that white Americans in the urban/ suburban Northeast accounted for a disproportionate amount of the total American magazine circulation, it seems likely a large number of middle- and working-class white Pittsburghers read *Reader's Digest* each month.[4] And if, as Joanne Sharp has argued, over half of the magazine's readers read each issue from cover to cover, it also seems likely that the periodical's framing of public-sector unions affected Pittsburghers' view of the teacher strike.[5]

The public discourse around the Pittsburgh strike that began in December 1975, then, drew heavily on the national context of public-sector unions (prominently featuring teachers) as well as a broad fear of becoming "another New York City." On December 9, *ABC Evening News*'s Howard Smith asserted on prime-time news that in the wake of the fiscal crisis many cities feared "New York–itis."[6] Pittsburgh was especially vulnerable to this fear because it had acquired a national reputation for fiscal prudence and represented a counter-model to Gotham's recklessness. An *ABC Evening News* story from June 5, 1975, for example, reported on the formation of MAC in New York, comparing the Big Apple to Pittsburgh. Whereas the "enormous" power of the largest American city's municipal unions inflated salaries—the report pointed specifically to the high salaries of public school teachers—and created superfluous jobs, Pittsburgh had remained "solidly in the black" for six years. Pittsburgh, the piece pointed out, actually *cut* public-sector jobs since 1970 and as a result had kept property taxes low.[7] Indeed, whether because the Steel City faced an attenuated flight of tax dollars to the suburbs compared to similar cities like Detroit, or because the State of Pennsylvania paid the full freight for the city's AFDC expenses, Pittsburgh's property taxes were indeed lower than any other city in the county and had not increased in six years.[8]

The Strike

Amid escalating fears about a New York City collapse, the contract between the PFT and the board of public education expired on November 30, 1975. In negotiations, the PFT sought pay raises over two years ranging from $2,700 at the bottom of the scale (for teachers making around $8,700 a year) to $6,000 for the highest-paid teachers (who at the time made $16,700). For comparison's sake, Pittsburgh's starting salaries were $500 lower than in Philadelphia, $900 less than in Detroit, and almost $2,000 less than in Chicago.[9] Pittsburgh teachers' salary demands would have put them above all of the others in this field, but they certainly did not expect to see their proposals fully realized, and they also expected inflation to continue to increase. The board could not raise taxes without permission from the state legislature and argued they would be irresponsible to agree to raises that were not paid for; instead, they offered an immediate $1,000 across-the-board raise with the second year subject to more negotiation. Teachers also wanted lower class sizes, and the board wanted to extend the length of the high school teaching day from five periods to six.[10]

On November 30, the teachers voted to strike. For several weeks, until the Pittsburgh Council of Parent-Teacher Associations sought a court injunction to "prove that the strike constitutes a public danger," there was little negotiation.[11] On January 3, Justice Alfred Ziegler of the county court of common pleas ordered the teachers back to work, and when they refused, held the union in contempt, seizing union assets to pay fines. Union leaders continued to ignore the order, and on January 12, Ziegler ordered a fact-finding panel to recommend a solution to the strike. The panel recommended the board offer more generous pay increases—though well short of what teachers wanted—and for both parties to submit to binding arbitration. Both refused: the board continued to argue it could not offer raises it could not pay for, and the PFT argued that binding arbitration might prevent them from salary increases that could keep up with inflation. As PFT president Fondy addressed a teachers' rally: "Let's see if I can put this simply. We are not going back to work until we have a new contract. Is that clear enough? No contract, no work!"[12] With all options exhausted—short of a court order to arrest teachers—the board settled by agreeing to raises of 11.3 percent in 1976, 8.3 percent in 1977, and 5.1 percent in 1978, bringing the lowest- and highest-level salaries to $10,000 and $20,000, respectively, by the end of the contract. On January 26, the teachers ratified the two-and-a-half-year deal, and classrooms across the city reopened the next day.[13] Pittsburgh teachers were now paid more than their counterparts in Philly, while school superintendent Jerry Olson warned there would be layoffs the following fall.[14]

The Public Discussion

A variety of voices contended to make their case about the strike's meaning. Aside from the board's austerity message, the mayor—Democrat Pete Flaherty—also argued for limiting expenditures so Pittsburgh could avoid becoming "another New York, or even slip[ping] to the brink of financial crisis like many other major cities in the United States."[15] Although some teachers wanted smaller class sizes, more support for their students, or school board policies to improve student discipline, most teachers' paramount concern was salaries that kept pace with inflation. As one teacher pointed out, from 1974 to 1976, the cost of living in Pittsburgh had increased roughly 30 percent while teachers had only received a pay increase of 3 percent per year.[16]

Pittsburgh, like many of the other cities in this study, featured racial conflict between teacher unions and the African American community. Many of these tensions came out during the 1975–76 strike. As in other eastern and midwestern cities, African Americans migrated to Pittsburgh during and after World War II, and both the federal government and real estate industry combined with federally backed urban development to segregate blacks in specific areas of the city—most prominently in the inner-city Hill District. The African American population of the Steel City increased from just over 80,000 in 1950 to 105,000 in 1970 (an increase from about 12 percent of the total population to 20 percent). The unemployment rate for blacks in the city in the 1950s and '60s was significantly higher than that for whites, and white Pittsburghers saw their salaries increase at a much faster rate. As in New York and Philadelphia, schools were largely segregated, and black students attended overcrowded schools in disproportionately older facilities. Black activists demanded more black teachers and principals, black history, and better facilities. In 1968, students shut down schools across the city, winning many of their demands.[17] Still, as late as 1970, the city only had about 400 black teachers (out of a total force of 3,300). This 13 percent was way out of proportion to the 40 percent of the student population that was black.[18]

In the early 1970s, the PFT became embroiled in racial conflict. In 1970, after the Pittsburgh teacher union had been certified as the teachers' bargaining agent, the union pushed the board of public education to establish eight special schools for "disruptive students"—at a cost of $1 million. That

May, the union's vice president even hinted that teachers might eventually strike to get the board to institute the new schools. The Pittsburgh NAACP vigorously opposed the proposal, pointing out that it would arbitrarily stigmatize black students.[19]

The issue only became more contentious. During the course of the 1970s, the school-age population of the city declined while the percentage of black students increased.[20] Indeed, even before the 1975–76 strike, the teacher union had sparred with the black community over school discipline. For much of 1975, the PFT pushed to restore corporal punishment to deal with increasing "disciplinary problems" at a number of predominantly African American schools in Pittsburgh. Teachers held a demonstration at the board of public education's headquarters in May 1975, and in the fall pushed the board to propose stricter discipline in the schools. Fondy, for instance, publicly argued that "unruliness has made it literally impossible to teach the kids in the public school classroom anything."[21] The PFT made noise about making discipline a major point in contract negotiations but chose mainly to emphasize salaries.

When the *New Pittsburgh Courier*—the city's African American newspaper—asked black Pittsburghers about the corporal punishment proposal, residents viewed it through their perceptions of a teaching force they believed had little enthusiasm for educating black children. Mother Arnet Tigney, for example, wouldn't "let anybody up at the school where my kids go hit them. Mostly because a lot of times there are teachers who don't like certain kids. . . . Usually the subject is selected early in the school year and the picked on Black kid or kids, whatever the case may be, will catch hell all school year long." Neither did parent Darlene Moore approve of "someone else beating on my kids. . . . They hire them so-called teachers to teach not to be beating on the kids." Joydel Johnson didn't "want any of them white teachers beatin' on my kids and I wouldn't stand for it."[22]

When the *Courier* asked black Pittsburghers about the possibility of a strike three weeks later (also inquiring whether city teachers should be required to live in Pittsburgh), several residents expressed their frustration with the union. Ken Rawlins believed "the teachers are making too much money already and they don't really need a raise, so they're not entitled to the right to strike. . . . Furthermore, the board should make laws which keep the teachers from living out of the city. It stands to reason that teachers who reside in the city can better understand the problems of the inner-city youth." Student Brenda Coleman argued that "the more money these teachers get the less that they want to teach us, so I don't think that they should be allowed to get a raise. They aren't teaching us anything new, so how do they deserve one? . . . Most of the time it seems that we go to school to learn, but we end up learning on our own."[23]

As in Philadelphia and New York City, even in the absence of a real fiscal crisis, the public discussion about the strike also focused on the illegal nature of teachers' actions after the injunction, as well as redefining political allegiance along an axis of productivity and linking "unproductive" unionized teachers with the students they taught. Indeed, after the strike became illegal and the strike dragged on, sympathy for teachers evaporated, and, in the growing narrative about the strike, teachers increasingly represented the culprit responsible for high taxes and the threat of fiscal precariousness in the Steel City.

The two major Pittsburgh dailies, going back to 1968, criticized teacher strikes (see chapter 1). Still, although the newspapers rebuked teachers in both 1968 and 1971 for walking out of classrooms, they had supported higher salaries back then. That had changed by 1975, even if their comments during the first days of the 1975–76 strike lacked histrionics. An editorial from the *Post-Gazette*, for example, hoped that "this one will be as short as the first two," but nevertheless asserted that "if it isn't, the skies won't fall." More harmful, according to the editor, than the shutdown of the schools was the "inflationary impact" of high costs and high taxes brought on by teacher salaries. It urged the board to be "fiscally responsible"—even more crucial in the wake of the New York City crisis—and to resist large increases in salaries.[24]

Early letters to the editor also criticized teachers for what they believed was an ill-advised walkout. Writing on November 29 to laud one of the newspaper's recently published cartoons, Mrs. E. W. Luttig asked the *Post-Gazette* to "reprint the picture and accompany it with a strong editorial." The newspaper obliged. The cartoon showed a suited man—presumably the superintendent—leaving the "Board of Education." A caption underneath read, "Shouldn't schoolteachers be paddled if they walk out on a strike?" Mrs. Luttig concluded by arguing that "the teachers' actions are the worst example possible. Can't somebody throw Fondy out?"[25] A letter from Craig Martin, who lived in a Pittsburgh suburb, in the next day's edition tied Fondy to a larger distrust of politicians in the Democratic-majority state assembly. Martin sarcastically asked:

Pittsburghers: Why don't you run Al Fondy for a seat in our Harrisburg legislature? . . . In the House or Senate Al could vote himself a raise every year or two without the fanfare of extended negotiations. Al could stay off the job for any number of sessions and no demands made. Al could vote himself any number of fringe benefits, none of which need to be negotiated or publicized. And, in addition, Pittsburghers, Al would be off your immediate backs. You might, in this way, share him with all of us Pennsylvanians.[26]

Clearly, Martin found both Fondy and the legislature equally guilty of violating the state's citizens.

Some commentators, while not exactly supporting the walkout, nonetheless supported raises for teachers—at least early in the strike. It seems likely that these observers understood that inflation affected teachers as much as it did taxpayers. Pittsburgher Betty Dunlap, for example, asked, "Why is the school board trying to deny the people who give our children their chance to be successful as adults . . . a decent living wage?" She argued that, without professional salaries, "our children will be spending five days a week with people who are teaching only because they can't do anything else."[27] A letter from parent Janice Kane on December 6 was more critical of the strike. Highlighting the broad acceptance of the notion that a coalition of the unproductive caused the fiscal crisis in the Big Apple, she believed that "Mr. Fondy and the PFT have not learned any lesson from New York City's near collapse from deficit spending largely incurred due to excessive demands for city employees coupled with expenditures necessary for swollen welfare rolls, generous free education facilities and other group pressure demands." If this rhetoric echoed language used by many New Yorkers as well as national media outlets like *Reader's Digest*, Kane nonetheless believed that "it is certainly important for any group or individual employee to air grievances and better working conditions and receive adequate compensation." Indeed, she thought the board should raise teacher salaries, but the question for her centered on tactics. Those teachers "who enter the field of public trust because of a desire to teach coupled with the ability to do so creatively and successfully" were no longer "the majority." The teachers who cared about their students, according to Kane, highlighting teachers' long association with care work, should cross the picket line and teach.[28]

In the second week of the strike, critics more firmly connected unionized teachers and the efficacy of liberal education policy. I. Hershorin's letter on December 8, for instance, upbraided teachers not only for striking but also for their poor performance. Identifying himself as the parent of a high school senior who used "fingers when doing math homework [while] telling me they were permitted to do it in school," he argued that "the basics of education have not deserved the caption 'quality education' for years. More money will not suddenly make better teachers."[29] Two days later, the *Post-Gazette* published a letter from Paul Mickey Toner, who also criticized the union rank and file. He asserted that teachers were "indifferent" in the classroom, and, in striking, were acting like the "petulant" children they taught—which they didn't do particularly well. "Today's student," according to Toner, was a "functional illiterate." On December 12, two commentators responded directly to Toner's attack. The first happily asserted that he "has distilled a series of confusing issues into focus." He wished Toner headed the PFT because, if so, "our teachers would be teaching." The second, O. F. Jedlick Jr., espoused a free-market nostrum that would gain wider purchase in the coming years: "Let ALL salaries be determined by the scholastic achievement of their students."[30]

As criticisms of the teachers increased, they became defensive. On December 9, for instance, Ruth Hertzberg responded to Kane's letter bemoaning teachers' supposed lagging commitment to their students. Hertzberg justified the strike by pointing out that the board's proposed $1,000 raise—accounting for inflation—would have actually represented a pay cut for most teachers. This presented a problem, she argued, because teachers provided opportunity through education. In Pittsburgh, where "the majority of students attend decrepit schools, with inferior facilities, and study from outdated textbooks," highly dedicated teachers—the ones who really cared for their students, in her view—brought the students' only hope for a "quality education." And the only way to maintain highly qualified teachers was to pay them well. By this logic, the "truly dedicated teachers" were on strike to ensure quality learning conditions for their students.[31]

An African American elementary school teacher and PFT member named Effie Moore admitted that, for the black community in Pittsburgh, the strike represented a "major catastrophe." She argued that black students in the district already found themselves at a disadvantage relative to many whites because "work for both black parents is an absolute necessity to make ends meet ninety-five percent of the time." The strike, for Moore, resulted not from "greedy" teachers, however, but from a damaged system of political economy. The choice was evident: "Quality education is costly from every angle. If the city of Pittsburgh really wants quality education for its youth, then citizens must be willing to pay the bill."[32]

Though not responding directly to Moore's letter, J. S. Burns, from the suburb of Allison Park, could have done so. Burns argued that the "just raise taxes" attitude clashed with the public interest. He reiterated the connection between Pittsburgh and Gotham, averring that the "fallacy" of raising taxes "is nowhere more evident than in New York City, which rewards its municipal employees in a more than princely manner, and has succeeded, in the process, in eroding a good share of its tax base, with more to follow."

He argued that Pittsburgh would be better off with an "austere" school budget.[33]

As the strike approached a month in duration, critics focused specifically on the teachers' violation of the helpless taxpaying "public." On December 30, for example, Don Meier, from Erie, Pennsylvania, explained that public-sector strikes differed from those in the private sector. Under "private enterprise," consumers could obtain services elsewhere when one company was struck. Not so in the public sector. Because "there are no alternative sources" for education, teachers had "violated" a "helpless" public.[34] Although Patty Calderone called the board "cheap," she directed most of her disapprobation toward "greedy" teachers who "will try to get everything they can, even if it means bankrupting the board of education."[35]

As was the case in other cities, the strike's criminalization opened up teachers to allegations that they violated law and order. On January 3, Judge Ziegler issued the injunction because of the strike's "general harm" to school students, the difficulties for parents who could not afford child care, and nonstriking school-district employees who had been laid off.[36] On January 6, both the *Press* and the *Post-Gazette* posted editorials denouncing the strikers for not returning to work and highlighting the continued failure of liberal public-sector labor policy. The *Post-Gazette*, for example, speaking to the obligations of teachers to nurture morals in future generations of Americans, asked that teachers "consult their consciences and realize the necessity of obeying the law." If they wouldn't, the editorial asked Ziegler to fine and/or jail union leaders.[37] The *Press* flatly asserted that the strikers had "declared themselves the only arbiters of what's legally right and what's legally wrong in deciding whether or not to obey a court order."[38]

In the *Post-Gazette* on January 8, two documents on the editorial page show the increasing rift between teachers and their critics. Cy Hungerford attacked the union with a cartoon titled "The Spirit of '76." The nation's bicentennial celebration was on the cartoonist's mind, and the graphic showed Albert Fondy dressed in Revolutionary War garb and flanked on either side by a drummer and a flautist. Fondy held a sign reading "We Will Fight to the Finish." The board of education, represented by an emasculated, bespectacled man prone on the ground, simply exclaimed, "Help!" The cartoon asserted that the union perverted the legacy of the American revolutionaries by victimizing the city's legitimate authorities. Below the cartoon, a letter to the editor from Jim Beyer, a Pittsburgh teacher, portrayed the strike in a vastly different light. He pointed to the high inflation of the previous two years and defended Fondy, noting that in acting "like any other labor leader," he defended his members' interests. Finally, Beyer upped the ante, averring that critics of the strike used the same arguments "that perpetuated the sweatshops of the 1920s."[39]

After Ziegler fined the union for contempt, public criticism of the teachers boiled over. Pittsburgher Robert McCully, for instance, asserted that Fondy "has taught impressionable youngsters to defy the judicial system in this country. . . . He has taught this community that if you don't get your own way, let the city be damned. The teachers may have a justifiable case in their own minds but breaking the law by defying a court injunction is certainly not the answer." Bob Howard, from Aliquippa, used similar language to characterize the strike: "I see Albert Fondy is at it again. He is willing to defy the law and the judge who is trying to enforce it."[40] While also emphasizing the "defiance" of the striking teachers, Hugh Young highlighted a deeper shift in the public reaction to the strike. Now, as the impasse between the board and the PFT mounted, and the teachers refused to relinquish their negotiating leverage, he called their demands "totally outrageous." A final letter writer, a mother from a Pittsburgh suburb, framed the choice in terms exhibiting what Joseph McCartin has shown in the 1970s to be the blueprint for Reagan-era labor relations, the "striker replacement strategy": "We mothers should stick together and protest too. Make [the teachers] work or fire them."[41]

Two days later, a letter from Judy Holzwarth to the *Pittsburgh Press* evidenced how far some Pittsburghers' views of teacher unions had shifted since the late 1960s. Holzwarth had been "in favor of the teachers getting the right to strike," but she had been wrong. She attributed her mistake to believing that although teachers deserved "the right to strike for decent wages, they promised not to hijack us for exorbitant wages once they received decent wages. They betrayed everyone who backed them then." Believing herself victimized, Holzwarth found herself unsure "if I really want some of the striking teachers back."[42]

Although much of the public discourse around the strike focused on the "defiance" of the teachers and their excessive demands, as the strike reached its crescendo, with Judge Ziegler seizing the PFT's assets and publicly contemplating jailing PFT leaders, commentary in the last two weeks of the strike opened up to larger debates about union rights in both the public and private sector and the larger fate of labor liberalism.

Robert Berkebile, for example, responded to an antistrike editorial in the *Press* and drew on Pittsburgh's labor history to defend the teachers' right to strike; indeed, he used the crisis to develop a larger argument about unions. According to Berkebile, teachers were no more influential to young people than "parents, grandparents, uncles, or aunts, or older brothers or sisters."

Further, these family role models "sometimes strike, picket and defy injunctions." Perhaps some readers, those who still remembered the agonizing efforts to organize the steel industry in the 1930s—or even the 116-day steel strike in 1959—could understand the teacher strike in the context of the broader labor movement.[43]

Also on January 10—though from a vastly different perspective—a letter to a suburban newspaper connected the strike to labor history. Writing from Fayette City, a town about thirty miles south of Pittsburgh, J. W. Smith was upset by the teachers' disobedience, and it inspired him to make "the voices of reason heard." He argued that "when large unions including public employees such as our firemen, police and teachers refuse to obey injunctions of our judges and even so-called decent elements of our society engage in disorder to gain their ends, this nation is indeed in a sorry state."[44]

In the *Post-Gazette* editorial page from January 12, several interlocutors believed the teachers' demands and the methods they used to attain them had become a synecdoche for the American economic and moral decline that went hand in hand in the 1970s.[45] Hungerford's cartoon that day showed a gigantic male dress shoe stepping on a representation of Judge Ziegler. Labeled "Teachers' Walkout," the shoe might have led readers to believe the union's defiance figuratively crushed the very embodiment of law and order. A letter to the editor on the same page, another from Hugh Young, compared Fondy's statement that it was necessary to "defy" the injunction in order to maintain "quality education" to "the statements made during the Vietnam War about the necessity of destroying Vietnamese villages in order to 'pacify' them." To Young and many other Pittsburghers, the tactics of the teachers, like the tortured logic of the Vietnam War (Saigon had fallen less than a year earlier), signified the nation's loss of moral values and the inability of the liberal state to solve problems—in either Southeast Asia or western Pennsylvania.[46]

The paper's editorial that day connected fears of moral crisis to the larger economic crisis of the 1970s. In connecting "productivity and stability" the editorial lauded some of the neoliberal reforms undertaken by corporations, financial interests, and free-market ideologues in the decade. Specifically, the editor commended the building trades unions for agreeing to "end costly work practices which impede productivity and drive up costs." It further held up the example to the PFT, urging teachers to consider that "as employment costs rise, they must be offset by greater productivity and/or fewer workers."[47] The implication here, when one considers just this one page in a Monday newspaper, is evident. Economic and moral crises were linked, and the way out in each instance was halting excessive public expenditure on the city's "unproductive" employees.

An editorial two days later in the *Post-Gazette* also framed the strike within a discussion of market-based approaches to education. Calling the teachers the "new priesthood," the editor argued that the teacher union had become too powerful because the citizens of Pittsburgh had allowed teachers to convince them they were uniquely qualified to teach: "Not just in Pittsburgh, of course, but throughout the nation the education establishment has fostered the absurd notion that teaching has become some arcane technology-cum-mysticism, the practitioners of which through specialized training have learned the secret of how to pass on knowledge. . . . [The emphasis] should be on how much someone knows about what he wishes to teach rather than on how many credits he's accumulated in supposedly learning how to teach it."[48] This "commonsense" approach to education, a subset of which included numerous conspiracy theories about the NEA and the AFT, would increasingly come to define the assault on public education in the 1970s and beyond.[49] And it would set the tone for future calls for policy innovations—like vouchers and charter schools—supposedly hamstrung by the antimarket interventions of unionized teachers.

By mid-January, as was the case in New York City toward the end of the crisis, critics were moved to action. Jack De Girolamo, on January 15, for example, composed a letter vociferously separating the strike from the Steel City's labor history. De Girolamo admitted that "during President Roosevelt's time in office when labor laws were being revolutionized, the public accepted unionism, compulsory collective bargaining, and the strike as countervailing forces to the power of industry." With a public-sector strike, however, "there is no force within any public group that can set itself up as a counterforce" except, according to De Girolamo, in a thinly veiled threat, "the will of the people expressed at the ballot box."[50] The next day, E. W. Maslak asked, "When are the city's taxpayers going to realize that it is time for them to 'come down hard' on the teachers, the Pittsburgh Federation of Teachers, and Albert Fondy? . . . Next to former President Nixon and his criminal cronies the PFT has done the most to undermine the values of the young people of our area."[51] By the strike's waning days, letters to the editor in suburban papers, too, exhibited a "taxpayer consciousness." In the January 24–25 weekend edition of the *North Hills News-Record*, for instance, David Wehner suggested that "taxpayers must have the legal right to contest any wage demands made by the teachers' union which might result in an unjustified tax burden."[52]

Teachers defended themselves even more vehemently. Dennis LaRue, a "holder of an expired teaching certificate," responded to the "New Priesthood" editorial of January 14. He asserted that "mastery of subject matter alone does not allow one to stand before a group of students, instantly the fount of all knowledge, and impart it to young people not especially interested in learning." Richard Price, a high school English teacher, believed that "this nation was founded by breaking unjust laws. . . . Great lawbreakers of the past include Thomas Jefferson, George Washington, Benjamin Franklin, and Thomas Paine. The list of Americans who have had to break unjust laws in the fields of labor, women's rights and civil rights to achieve justice is long and honorable."[53]

The lines between teachers and "the public" had been drawn quite starkly, and teacher protests only further inflamed opponents convinced that teachers' violation of the law had enabled a lapse in morals and a breakdown in the ideal of law and order. Teresa McNulty, for example, passionately responded to Price's letter with a question: "How dare he place the Pittsburgh teachers' demand for more money on a par with . . . 'other lawbreakers'?" She also asked Price to "stop painting pictures of teachers as latterday founding fathers. The public is not deceived." In addition, C. Gaetano responded to Price by arguing, "There is only one place for teachers of Pittsburgh—not behind the bargaining table but behind the desk."[54]

The Aftermath

As crises often do, the strike forced many Pittsburghers to confront their assumptions about the nation's basic political divisions and who they believed the labor-liberal state served. A series of summaries by local newspapers near the strike's conclusion expressed both fear and urgency. The headline of an editorial in the *Leader Times* of Kitanning on January 21, for instance, announced that "teacher strikes underscore growing public helplessness."[55] The *Pittsburgh Press* bemoaned the abject failure of PA 195 to prevent public-sector strikes, asserting that "with the public-employee unions powerfully armed, and often heedless of anybody else's interests, the public needs more adequate defenses. Clearly, the weapons now available to the public don't have much more force than a popgun."[56]

In addition, the public discussion of the Pittsburgh teacher strike highlights a problem with some scholarly attempts to understand the conservative turn by many white working-class Americans in the 1970s. In his description of the 1972 presidential election, for example, Jefferson Cowie argues that Nixon turned blue-collar workers from the New Deal coalition by emphasizing cultural values—like support for the Vietnam War and "law and order"—at the expense of worker-friendly economic policy.[57] To many white working-class Americans, however, cultural values like respect for law and order and hard work were intimately tied to economic concerns. Furthermore, in a time of stagnating wages and inflation, many believed that curtailing the liberal state—at least those aspects like teachers' demands for higher salaries and black parents' for better schools—made good economic as well as cultural sense. As a result of the history of conflict between teacher unions and the black community, however, it was more difficult to develop a coordinated response to a new coalition of corporate elites and producerist whites.

And yet, the shifting terrain of American political allegiances still flew partially under the radar. Just two weeks before the Pittsburgh strike ended, a reporter from the *Press* asked I. W. Abel, president of the United Steel Workers of America, to predict what the labor movement might bring in the next century. Abel admitted that "labor is experiencing some backlash from the public now . . . mostly because of strikes by . . . police and teachers," but he ultimately concluded that "when you consider what we've gained during the past 40 years, it boggles the mind to think where organized labor will be in the year 2076." This statement proved to be a major miscalculation of the labor-liberal coalition's future; indeed, Abel missed the fracture taking place before his very eyes in the Steel City.[58]

Perhaps the best example of the shifting position of political fault lines, however, is that the 1975–76 strike also inspired, as had the 1968 strike, a series of legislative hearings. By 1977, however, the optimism of 1970 that the rational administration of a well-constructed law could create labor harmony in the public sector had dissipated. New tensions brought on in large part by the Pittsburgh strike led, at the end of 1976, to the formation of a new Governor's Study Commission on the problem of public-sector labor law. Appointed this time by Democratic governor Milton Schapp, the commission held public hearings in Philadelphia, Pittsburgh, Erie, Harrisburg, Allentown, and Scranton during the spring and summer of 1977, and deliberations continued into 1978. Although the governor intended the hearings to investigate all public-sector unions under the jurisdiction of Pennsylvania Act 195, they clearly revolved around teachers. Representatives from teacher unions helped to prevent the commission from recommending newly restrictive legislation, but the battle lines that emerged in the hearings evidenced a growing suspicion and hostility toward public-sector unions. In particular, critics of teacher strikes highlighted the failure of the state to solve the labor problem and prominently focused on violations of law and order.

James Scott—president of Pennsylvanians for the Right to Work—was one of the first to testify at the hearings. Scott's group represented the state affiliate of the National Right to Work Committee (NRTWC), a political advocacy organization primarily concerned with halting "compulsory" union membership in states across the United States. In his testimony, Scott emphasized, as the NRTWC had done since the late 1960s, that public-sector unions inherently violated local communities.[59] Indeed, Scott argued that "particularly hard-hit by public employee strikes have been the students and taxpaying parents in our public school systems." He proposed a new policy outlawing agency fee arrangements (in which all public employees in a bargaining unit were obligated to contribute to union representation costs), as it would weaken unions so that sustaining strikes would be more difficult.[60]

In the Philadelphia hearings, W. Thacher Longstreth, president of the Philadelphia Chamber of Commerce and Republican mayoral candidate for the City of Brotherly Love in 1972, described the impact that strikes had had there (Longstreth had been on the city council during the 1970 strike): "The disasters visited upon community after community by striking public employees is eating away at the very fabric of society and hurting everyone." Although strikes might be okay in the private sector, testified Longstreth, because "we are talking about a strong union but no counterpart in strong management," strikes in the public sector should be outlawed.[61] Here Longstreth turned on its head the notion that labor could maintain social peace by serving as a countervailing force to corporate power, arguing that the public could not viably curtail public-sector union influence.

School-district representatives, unsurprisingly, criticized teacher strikes that had occurred under PA 195. As might be expected, some focused on unbalanced budgets and union attempts to control working conditions; a larger portion of the criticism, however, centered on the illegal nature of strikes and the bad example they set for students. A school board president in a rural district forty miles outside of Pittsburgh described a teacher strike the previous winter; he believed it fit into a larger story of American moral declension: "There were tears in my eyes, almost freezing on my face to think that our education system had come to such a point that teachers were trying to stop students from going to school." Jerry Olson—Pittsburgh superintendent in 1975–76—argued that the most important detriment to the city's schoolchildren was not lost instruction time. He claimed that "whenever teachers defy a return to work court order, their actions confuse students as to the need to respect the law and to heed the word of the court. This in fact is a far greater damage than school days lost."[62]

Parent groups also criticized teacher strikes and advocated more restrictions on unions. In her testimony, Philadelphia Home and School council president Edna Irving asserted that strikes like the conflict in 1972–73 brought a "breakdown in respect" for Philly's young people. The Home and School council, she said, believed that "people who are involved in public education are in the same categories as our policemen, our firemen, the people who provide our medical services. There should be no breakdown in services."[63]

Bob McCannon, president of a "grassroots" group in suburban Philadelphia formed specifically to fight teacher unions, stated unequivocally that "Act 195 does not reflect the public interest." He wanted the commission to recommend restricting strike rights and forcing union votes to be by secret ballot. McCannon chalked up teacher strikes where he lived to "mob rule" and believed teachers should act like "professionals" instead of union members.[64]

The shift in the political discourse in Pennsylvania about teacher strikes between 1970 and 1977 signified, in a fundamental way, two major developments. First, there was a growing skepticism about the ability of rational policy to curb the turmoil resulting from teacher strikes. The 1977 commission—comprising several labor negotiators, attorneys, and an academic—did not recommend radical revisions to the law, instead advising Schapp that the bargaining process should begin sooner and school districts should not be penalized by the state if they could not guarantee 180 school days.[65] No one, however, in 1977 testified at the hearings (or wrote in a newspaper editorial) that PA 195 had been the "wondrous new thing" that the *Harrisburg Patriot* in 1970 thought it might be.

Second, by 1977, a much stronger opposition to public-sector unions—particularly teacher unions—had clearly emerged. There was nothing inevitable about this development; in his testimony to the commission, Pittsburgh Federation of Teachers president Fondy argued that teacher strikes often lasted so long because school boards simply waited for an injunction in order to gain leverage. Without the "public welfare" clause in PA 195, he suggested, strikes might have ended earlier because boards of education would have been less likely to drag their feet in the hope of legal action.[66] One could certainly view this claim as self-serving, but the larger point here is not whether the position of the board or the union was justified in Pittsburgh in 1975–76. Although we might very well hold a mostly white teaching force accountable for not fighting hard enough to ensure all students could learn in Pittsburgh, it is difficult to blame teachers for striking for salaries to keep pace with inflation, and it is also difficult to blame a school board for not wanting to spend more money when increasing taxes was less and less palatable. In any event, precisely because of the intractability of the problem, the phenomenon of public-sector strikes—especially by teachers—represents a key component of a deeper political shift in the 1970s. The Pittsburgh strike shows how the earlier conflicts over public-sector labor law and struggles for racial equality in the schools combined with the new terrain on which American cities operated following the New York City fiscal crisis to further rip apart the labor-liberal coalition.

CHAPTER 6

THE "FED-UP TAXPAYER"
St. Louis, Philadelphia, and the Eclipse
of the Labor-Liberal Coalition

The events of 1978, taken together, crystallize many of the big ruptures of the 1970s. In February, the National Organization for Women declared a "state of emergency" as the right-wing backlash against the Equal Rights Amendment left it stalled a year out of the original 1979 ratification deadline. In April, the U.S. Senate ratified a treaty to return the Panama Canal to Panamanian sovereignty, symbolizing many Americans' fears about the waning of the nation's power. In early June, a bill to more stringently enforce the National Labor Relations Act against increasingly recalcitrant corporate employers fell two votes short in the Senate, underscoring the concerted effort of big business and free-market ideologues to dictate national policy. Days later, California voters passed Proposition 13 to dramatically limit property taxes, apotheosizing a decade of antitax animus in the Golden State and igniting similar efforts elsewhere. In September, the Supreme Court limited the scope of affirmative-action programs in its decision in *Regents of the University of California v. Bakke,* coming on the heels of a virulent reaction among many whites against the historical understanding of the connection between race and economic opportunity in the United States. Counsel for the aggrieved party, Allan Bakke, argued that positions in the University of California–Davis's medical school set aside for historically disadvantaged minorities victimized him.

While the events may have been highly dramatic, the trends—fear of American decline, backlash against structural understandings of inequality, and corporate insurgency—had begun earlier, at least as early as 1968. Still, these big events symbolized the new trajectory of the nation's political center: a deepening sense that the labor-liberal state now victimized both white middle-class Americans and corporate America, which led to a host of new calls to let the market structure life in America.

This chapter returns to Philadelphia and St. Louis, and shows how teacher unions there operated in this new political atmosphere. In simultaneous strikes in early 1973, the discourse in each city had moved from a robust discussion about urban political economy toward one that, with strikes criminalized, featured a profound pessimism about the state's ability to control its labor force. In 1979, St. Louis teachers struck again, and in both 1980 and 1981, Philadelphia teachers walked out. The national context, however, had shifted dramatically—in part because of the intense conflict over education in the nation's big cities during the late 1960s and '70s—and this chapter demonstrates the opposition teacher unions faced in this new era. In St. Louis, it took a bare-knuckle two-month strike for teachers to gain a very small salary increase and no significant change in what they argued were excessive class sizes, while in Philadelphia, the teachers faced concession bargaining as a Democratic mayor forced the city on a path to austerity. The public conversations in each instance are telling, particularly in contrast to 1972–73. In both St. Louis and Philly, teachers found themselves with few allies. Indeed, many critics argued that there was little reason to even attempt to salvage the inner-city public school system—with, in their view, its unproductive teachers and hopeless students—in any capacity.

Public-sector strikes by nature revolve around public policy because the salaries and working conditions of government workers such as teachers inherently involve interventions by the state.[1] This was particularly so in the brutally bitter Philadelphia strike in 1981. Indeed, the emergence of arguments favoring market-based privatization—from everything to just allowing the public school system to die outright to school voucher programs—symbolized the new degree of pessimism toward the efficacy of the liberal state that increasingly marked American political culture by the early 1980s.

The St. Louis Strike

Strikes declined after the high-water mark of 1975–76, but they did not disappear.[2] In the fall of 1977, for instance, Oakland, California, teachers walked out for a week and a half, winning a single-digit salary increase and smaller class sizes.[3] AFT teachers in Lakeland (Westchester County, New York) went on strike for forty-one days, the longest strike in the state's history to that point. Although teachers there received a 13 percent pay increase over three years, six union leaders served a month in jail in what a reporter for the *New York Times* called "a dramatic taxpayer reaction to increasing education costs."[4] April 1978 brought a twenty-three-day strike in Toledo, where the schools operated with a "bare bones" budget after citizens continued to turn down tax increases. Teachers ended up winning a 13 percent salary increase over three years.[5]

In the fall of 1978, teachers in Wilmington, Delaware, went on strike for five weeks. That conflict stemmed from a court-ordered desegregation plan that merged the mostly black Wilmington School District with several white suburban districts. Wilmington teachers won hard-fought pay increases in 1975 (see chapter 4), and the suburban teachers wanted to be paid as much as their city counterparts. About 1,000 Wilmington school employees—nearly all black—opposed the strike. The 1978 strike was settled by freezing the higher-paid teachers' salaries until the suburban teachers could be "leveled up."[6] A twelve-week strike in Levittown, New York, put a teacher union president in jail for three weeks.[7] In early January 1979, another strike in Newark was averted only after a New Jersey judge ordered the Newark Board of Education to restore jobs to laid-off teachers.[8]

As in Newark, St. Louis teachers also faced a conflict revolving prominently around fiscal limits. Indeed, January 1979 was not a propitious time for the St. Louis Teachers Union's (STLTU–AFT Local 420) contract to expire. The problems St. Louis faced in 1973—a declining tax base and a reluctance on the part of the State of Missouri to offer additional help to the school system—had not diminished in the four years after the STLTU won an election as the teachers' exclusive bargaining agent in 1974. In fact, the State Department of Elementary and Secondary Education downgraded the city schools' AAA rating (the highest possible) in 1975 because of excessively large class sizes and not enough preparation periods for elementary school teachers.

The teachers also had to contend with the timing of California's Proposition 13, an amendment to that state's constitution capping local property-tax assessments and setting future limits on tax increases. The ballot initiative, passed by a margin of 65 percent to 35 percent, was borne of legitimate grievances among many California homeowners as inflation and skyrocketing housing prices combined with a $4 billion state budget surplus to make high property taxes seem both unnecessary and capricious. The debate over Prop 13 was intimately tied to both public employees—the state's public-sector unions mobilized to defeat the initiative—and the state's education system because local school districts directly depended on the property tax for funding. Although the state used its surplus to subsidize local districts, schools represented the public service that was most affected by the massively reduced revenue. Perhaps more important, however, as Robert Kuttner argues, "a Proposition 13 mythology grew up overnight. National political commentators . . . decided in chorus that June 6, 1978, marked the day American voters put a limit on the rising size and cost of government."[9]

Former governor Reagan, for example, in a nationally syndicated column appearing in the *St. Louis Globe-Democrat*, believed the "tax revolt fires" lit by Prop 13 had created a new "political wind" in which Americans wanted more accountability from a more limited government.[10]

Legislators, political pundits, and antitax activists in other states—including Missouri and Pennsylvania—immediately considered similar tax-limitation proposals, through either legislation or ballot initiative. Indeed, the overwhelming margin by which Prop 13 passed put taxes at the center of political debate in Missouri and the rest of the nation.[11] Although it would not happen until 1980, Missouri passed its own version of referendum-based tax limitation when a businessman named Mel Hancock led a petition effort to limit taxes. The constitutional amendment, which pegged revenues to the income of the state's residents, passed by ten points the same Tuesday in November that Reagan won the presidency.

Furthermore, on the heels of the filibuster of labor law reform the previous summer, Missourians for the Right to Work succeeded in placing an initiative on the ballot to make the Show Me State right-to-work in November 1978. Buttressed by a strong union turnout, the measure failed by about twenty points, but the fact that it was even on the ballot evidenced the belief on the right that attacks on unions could succeed. Next, in late November, a strike by several newspaper workers' unions shut down both of the city's major newspapers for almost two months, leaving citizens with a print news blackout. The strike ended only after pressmen gave up overtime shifts and allowed the newspaper to cut jobs through attrition—steps the newspaper argued it needed to compete in a market increasingly dominated by television news. Then, in January, the teachers' contract expired as Great Britain faced the most critical moments in the 1978–79 Winter of Discontent that would ultimately take down the Labour government and bring neoliberal Margaret Thatcher to power. The very day after newspapers returned to production in St. Louis, those reading the *Post-Dispatch* could see coverage of a United Kingdom delving "deeper into chaos" as "millions of workers faced layoffs" and transportation strikes left stores bereft of basic food and other necessaries.[12]

Finally, the teachers' contract expired just after the official revelation of the inflation numbers for 1978. The inflation rate for that year was 9 percent—highest since the double-digit inflation of 1974—and President Jimmy Carter's 1979 "austerity" budget, announced just as the teachers' strike began, proposed substantial cuts to federal spending on social programs, public-service jobs, higher education, and school lunches, ratifying the notion that public-sector spending was tied to high inflation rates.[13]

In this context, 4,400 teachers and other public school employees represented by Local 420 voted to strike on January 15, 1979.[14] The teachers struck with two goals in mind: salary increases to keep pace with inflation, and, to regain its AAA rating, smaller class sizes and daily preparation periods for the district's elementary school teachers. STLTU president Evelyn Battle, speaking for the 80 percent of St. Louis teachers who voted to strike, encapsulated the frustration of a decade in which compensation lagged and working conditions barely changed: "We've waited and waited, but now our members feel that this is the time to take a stand. I feel sad it had to come to this, but I'm excited that teachers at last have said, 'we're tired of being forgotten.'"[15] The board of education, citing lack of money, offered a $200-a-year across-the-board raise (from a first-year teacher's salary of $9,650) and two prep periods a week for elementary teachers. Union negotiators argued that the board had underestimated surpluses for years (a position confirmed by analysis from local newspapers), and that they could afford a $1,000-a-year teacher increase, one prep period a day for all elementary school teachers, and lower class sizes in every grade.[16] Indeed, each side drew a line in the sand, and the lines were far apart: the school system wanted to maintain conditions more or less as they were, and the teachers wanted to be paid like their counterparts in bigger cities like Chicago, New York, and Philadelphia. Both board and teachers settled in for a lengthy stalemate.

The legal status of public-sector unions in Missouri complicated the negotiations. The state had not revised its labor law in the 1960s as had many eastern and midwestern states, and its cities worked with legal precedent to determine their policies. The most important case—from 1947—about public-sector bargaining in Missouri stated that public employees did not enjoy the same rights as private-sector workers under the state's 1945 constitution. Therefore, unlike in other highly unionized states like Pennsylvania and New York, school boards did not have to recognize collective bargaining. Further, strikes by public employees *were* explicitly forbidden. At the behest of the school board, circuit court judge Ivan Lee Holt Jr. issued a temporary restraining order the day the strike began to prohibit union leaders from encouraging the walkout. STLTU president Battle immediately disobeyed the order. When the board sought contempt citations, union leaders believed that they might face a crackdown by local authorities, and Battle proclaimed that "everyone must be willing to go to jail and stay there to show the board we cannot be forced to go back to work."[17]

As in strikes elsewhere, teachers defied the law, but the local judiciary, surprisingly, seemed to understand the futility of legal repression: Judge Holt refused to issue arrest warrants. Missouri—particularly St. Louis—had a long history of union culture, and pushing too far could have had political consequences, particularly when the union had the strong backing of the St. Louis Labor Council, whose leaders argued that the teacher union "is justified in its concerns for the quality of education our young people are receiving. We believe further that the economic demands made by the Teachers are also justified by comparison with other school systems of similar size."[18] Furthermore, as strikes in Newark and Philadelphia earlier in the decade had shown, mass arrests elsewhere had only served to heighten strikers' resolve.

Even the board backed off two weeks into the strike, dropping its restraining order and the back-to-work injunction under the court's consideration. The board saw this move as conciliatory and hoped the teachers would return to work. The teachers, who derived no practical benefit from the removal of legal barriers they would defy anyway, saw no reason to go back without a contract.[19] The board's "concession," however, represented a major public-relations victory because many opinion makers—from newspaper editors to individual citizens—believed that the teachers were being unreasonable after the board dropped its strategy of using the law to force teachers back.

In contrast to Republican governor Kit Bond in 1973, Democratic governor Joseph Teasdale took an interest in the strike, stating on February 24 that "I am prepared to use my full power as governor to assist in resolving this matter."[20] The strike would end, in fact, after Teasdale made almost $1.5 million available to improve the board's offer in early May, and the school board, in conjunction with parents' groups, reinitiated injunction proceedings against the STLTU.[21] The resulting agreement, reached two days after Holt issued a new injunction, hardly represented an overwhelming victory for the union. Teachers received a $1,250 across-the-board increase over the life of the two-year contract, which was marginally closer to the union figure than the board's original offer, with $1.4 million coming from Missouri and about $600,000 in funds raised from local businesses. Still, after a fifty-six-day strike, many teachers had hoped for a reduction in class sizes and more than two prep periods a week for elementary school teachers, neither of which had received traction in negotiations.[22] It was a win, but one in which neither working conditions had changed nor had the city's ability to pay desirable salaries through any source of permanent funding. In addition, the perception of the teachers and the school system would shift as the public discourse around the strike re-centered the identity of the public stakeholder from a civic recipient of education to the productive "taxpayer" who stood in direct contraposition to those she or he paid for the teachers' services.

From "Public" to "Taxpayer"

The major archival material for understanding the 1979 teachers strike in St. Louis is the editorial pages of the city's two major newspapers. Although both the *Globe-Democrat* and the *Post-Dispatch*—each of which opposed the strike for different reasons—had their own interests, each also attempted to show the paper's fairness by providing a diversity of viewpoints in the copious letters to the editor they published. Teachers, therefore, defended themselves in print; nonetheless, as the conversation developed over the two months of the strike, other views of the teachers increasingly characterized them as unproductive employees and asked whether they provided "value" to the "taxpayers" rather than viewing them as providers of a necessary public service to the city's youth (as in 1973).

When the strike began, the *Post-Dispatch* and *Globe-Democrat* espoused very different views of the teachers' position. The *Post-Dispatch*'s view remained basically as it had in 1973: teachers should be able to collectively bargain, but they should not strike, and the current action, though illegal, could be solved through compromise. In its first comment after the strike began, the *Post-Dispatch* asserted that "teachers deserve not only the recognized legal right to bargain but also legal provisions that can aid them when talks break down." Although these changes were nowhere near the horizon, the editor believed that "until such changes are made, the teachers ought to rely on gaining public support to pressure the board."[23]

The *Globe-Democrat*, on the other hand, had moved even further to the right in the six-year interregnum between strikes. Reflecting the insurgency of right-wing politics in the late 1970s, the newspaper increasingly filled its editorial pages with syndicated columns by conservatives like George Will, Phyllis Schlafly, Patrick Buchanan, Ronald Reagan, and James Kilpatrick. The *Globe-Democrat* had also jubilantly celebrated the victory of Proposition 13 the previous summer by running—among other pieces—an editorial on June 8, 1978, declaring that "Californians Kill 'the Monster'" and a cartoon several days later titled "It's the Aftermath of the California Earthquake," which showed the Washington Monument collapsing onto two fleeing men branded "Liberal Spending."[24]

It was no surprise, then, that the *Globe-Democrat* characterized the STLTU strike as "unconscionable" and asserted that in the future the board of education should insist on a "no-strike pledge as a condition of employment"—a redundancy given that teachers could already be fired for striking. The editorial page also dismissed teachers' concerns about salaries and the state of the schools by comparing them to silly, impudent undergraduates: "The striking

St. Louis teachers need to learn that they are not engaging in a fraternity prank or sorority stunt by being on the picket line. They are breaking the law and seriously disturbing the peace of the entire city." This interpretation of the strike was belied by the reality that this union of mostly female white-collar professionals walked off the job and lost salaries while risking jail time; *Animal House* frat mischief it was not.[25]

Teachers criticized the positions of both newspapers, justifying the strike in the context of the post-tax-revolt era that had put public employees on the defensive. Sue Pratt, a teacher from a St. Louis suburb, lashed out at the *Globe-Democrat*, for example, arguing that teachers only broke a law that was "unjust, discriminatory, and morally wrong. Other workers are allowed to strike and achieve cost of living raises which they would otherwise never receive." Pratt also criticized the *Post-Dispatch*'s view that teachers should try to change the law instead of striking: "Privileged groups seldom give up their privileges voluntarily, taxpayers seldom pay more taxes voluntarily. When all else fails, an issue must be dramatized through crisis and nonviolent action so that it can no longer be ignored."[26] A St. Louis teacher named Thomas Thavorides argued that the decision to strike represented a "very painful" choice, and teachers wanted to return to their classrooms. "Before we do," he continued, "we ask only that those schools be made worthy of the state's AAA rating and that we receive a salary commensurate with our education, responsibilities, experience, and efforts." Pointing to the board's practice of unilaterally adding extra "steps" to the salary scale, Wayne Gower averred that teachers were "not only the victims of inflation, but they are victimized and exploited economically by the same board of education to which they have dedicated their service."[27]

Commentary from the nonteaching public was limited during the first two weeks of the strike. After the board dropped its restraining order and still teachers stayed out, both newspapers criticized them even more vigorously. More important, however, each helped in its own way to reframe the strike around the rights of the productive taxpayer. On January 30, for instance, the *Post-Dispatch* again pointed out that teachers should be able to collectively bargain but also asserted that "the School Board, after all, is but the agent of the taxpayers."[28] Such framing defined the "taxpayers"—as the employers who paid for salaries—as the teachers' adversaries. As might be expected, the *Globe-Democrat* criticized the board's "conciliatory" gesture. Focusing on the illicit action, the newspaper argued that "the strike, being illegal, is indefensible, but the school board backed away from seeking a permanent injunction." More important, however, the editorial asserted that "the schools are not the private property of the school board or the teachers. The schools belong to the taxpaying people of St. Louis."[29] These two views of the strike reflected the new climate that had emerged by the late 1970s: the disaggregated "taxpayer" had become the most important subject in the dispute—not the students or their parents, and certainly not the teachers.

Public commentators exhibited their own taxpayer consciousness. An "Unsympathetic Taxpayer" criticized Wayne Gower's assessment of the district salary schedule and argued for a market-based structure. Identifying himself as a "practicing accountant for the past thirty years," he believed that "teachers have to be the only professional group that I know of who have the privilege of knowing in advance what their earnings will be. Other professionals, such as my group, must rely on the employer's yearly decision to measure the job we have done over the past year."[30] In a letter to the *Post-Dispatch* on February 8, "A Reader" argued that more than just parents held a stake in the strike: "Those of us who do not have children in school, but still do our duty supporting schools through our tax payments, are getting tired of providing teachers for babysitting each day. We are also getting tired of wasting tax money for labor relations specialists, labor attorneys, and the use of federal mediators." This comment, indeed, speaks directly to how many Americans had reassessed their relationship to government services like education: the "productive" taxpayer fatigued by paying for schools that did little more than "babysit"—gendered terminology for teaching that diminished its value—and for the salaries of an intelligentsia of labor experts and government bureaucrats.[31]

These circumstances forced teachers to defend themselves in the terms of the new taxpayer consciousness: as inherent adversaries. One teacher, for instance, lectured the "Unsympathetic Taxpayer" that his comparison of public- and private-sector jobs highlighted "misplaced" priorities. Attacking the notion that "taxpayers" could pay teachers meager salaries while criticizing them for a subpar school system, she placed the responsibility on taxpayers, asking, "How many times in the past has a school bond issue been rejected? The answer to that query would indicate to me that there is a profound lack of interest and concern for the children, the teachers, and the whole school system."[32]

As the strike reached six weeks by the end of February, taxpayer critiques became more pointed, focusing on the teachers' supposed lack of accountability. A "reader," for instance, argued that "keeping schools closed with an illegal strike is not an effective way to educate the system's 73,000 pupils. I can see why the citizens of St. Louis do not support tax levies for the schools. What kind of excellent school system is on strike when it should be providing the children of the city with quality education?"[33] A "Fed-up Taxpayer" wrote a scathing letter to the *Post-Dispatch* on February 28, criticizing teachers' excessive salaries. After an hourly breakdown based strictly on the amount of time teachers directly instructed students, the taxpayer suggested that readers "compare their own hourly rate with these, and decide themselves if these teachers are really underpaid. . . . And yet the union even says 'no' to a suggested performance audit to gauge how well the teachers . . . are doing their jobs; are they unwilling to face these facts also?"[34]

Kendall Wentz questioned whether teachers provided any value to taxpayers. While admitting they earned less than New York City garbage collectors, he still wondered whether teachers even deserved their present salaries: "Is the laborer worthy of his hire? How is St. Louis ranking on the Student Achievement Test Scores that are administered by the colleges?" He also asked whether the "gain in achievement [is] worth the expenditure of additional funds just for the sake of smaller classes?"[35]

Teachers' rhetoric became more defensive in the face of these criticisms, but moving onto the terrain of market logic undercut the notion that teachers contributed to a public good instead of representing just another commodity. Roger Faber, for example, offered a sharp rebuke to those who criticized the teachers by invoking the larger trajectory of labor politics in the United States: "Now don't remind me that I knew what I was getting into when I became a teacher. Take that philosophy and apply it to any wage earner who strikes for higher wages. Why weren't the paper handlers told that during their recent strike? Or, just try telling the teamsters, 'You knew what you were getting into when you became a trucker.'" With regard to the teachers' lack of consideration for their students, Faber admitted that "I like to see [teenagers] grow and develop into intelligent, well-equipped adults because I was their teacher. But my banker still wants money when my house payment is due and there are no grocers who will accept a pound of dedication for a pound of beef. . . . Believe me, parents," Faber concluded, "I like my job and I like your kids but I'll be darned if I will continually sacrifice the welfare of my own family for either of them."[36]

Robert Naumann responded directly to the "fed-up taxpayer's" characterization of overpaid teachers: "The public simply doesn't understand the teacher's job. No teacher can get by on a six-hour day. Lesson plans must be prepared. Tests must be administered and corrected. . . . This cannot be done, along with instructing the student, discussing materials and dealing with the many distractions that occur during the six hours the children are at school." Naumann also put the value of student learning in the terms of taxpayer consciousness, asking parents, "What value do you as a parent place on the healthy development of [your child's] mind? Remember, in this as in most other things, you generally get what you pay for."[37] Finally, Maureen Raucher reminded readers of the *Post-Dispatch* about the structural impediments to teaching at-risk students—an issue that had been front and center in the 1973 strike but which was virtually absent in 1979. "It seems as if all of society's ills," she argued, "are being placed on the shoulders of the teacher. . . . We're constantly waging the war on poverty, ignorance, and parental neglect. Is it any wonder that several recent studies have shown that inner-city teachers are displaying some of the same symptoms displayed by soldiers returning from war?"[38]

The strike clearly embittered both teachers and many members of a public who viewed their political agency as that of a taxpayer. E. M. Lessingham wrote to the *Globe-Democrat*, for example, that the teachers "joined a union organization and in the name of that organization held hostage for two months public school property. . . . I am outraged as a citizen, and I will never vote for a school levy tax again." Criticizing both the board and the union, John Smith pointed out that there would have to be "a power play to extract a school tax increase from the voters next year. . . . If the taxpayers figure this is even a remote possibility there will never be a 'majority' found again for a school levy!" Finally, M. Harris was "still burned up about the St. Louis teachers' strike. . . . If you ask me, St. Louisans should refuse to pay their property taxes for an amount of time equivalent to the teachers' strike."[39]

There was nothing preordained about these responses. There are many ways that outraged citizens can respond politically; these three commentators could have pushed for different school board representatives, stronger collective-bargaining laws that guaranteed full strike rights, or increased state or federal expenditures for beleaguered cities. These responses, however, indicated that after a decade in which contentious strikes had placed teacher agency at the center of the debate, criticism of high taxes had moved toward the center of American politics, and because the rupture stressed the victimization of "productive" taxpayers, threatening to withhold taxes made sense as a political strategy.

A temporary infusion of outside money narrowed the chasm between the positions of the board and STLTU, settling the strike after two months of negotiations. The gulf between what teachers thought they were doing and how the public viewed them, however, had continued to widen. Indeed, by the end of the 1970s, the basic axis of political debate during teacher strikes was fiscal—how much would the "taxpayers" be forced to pay for education, and was it worth it? The next two years would bring teacher strikes in Philadelphia, and these would revolve even more around the cost in taxes for teachers' salaries—and some critics would propose to maximize value through market-based models of education.

Two Strikes: Philadelphia, 1980 and 1981

Lengthy teacher strikes continued throughout 1979. As the St. Louis strike concluded in March, teachers in the nation's capital walked out for more than three weeks. About half of DC's teachers—members of the Washington Teachers Union (WTU–AFT)—went on strike illegally after the board of education refused to extend the previous contract, believing it gave too much authority to teachers over grades, discipline, and other educational policies. Unique in the late 1970s in that money was not the central issue, the strike ended when a DC judge ordered teachers back to work under the old contract.[40] When schools opened in the fall of 1979, there was a strike wave that was the nation's most extensive since 1975. By late August, teachers were on strike in Oklahoma, Louisiana, Indiana, Illinois, Ohio, Michigan, Missouri, and Pennsylvania. Between 50 and 70 percent of Oklahoma City's 2,200 teachers walked out for higher pay; the school board doubled the daily wage of substitute teachers willing to cross the picket line.[41] In Spokane, Washington, 1,400 NEA teachers defied an injunction against their illegal strike, as did 3,600 teachers in San Francisco; 3,350 teachers in Indianapolis went on strike, too, and Detroit teachers walked out again for three weeks.[42]

But the most significant strike that fall was in Cleveland, where teachers remained out of classrooms from mid-October until early January 1980. As in other major American cities, deindustrialization had severely eroded Cleveland's tax base—one official estimate put the decrease at $200 million from 1966 to 1976—and property owners resisted tax increases (which in Ohio had to be approved by the city's voters). By November 1977, the board of education struggled to pay teacher salaries.[43] In the spring of 1978, a federal judge ordered busing to desegregate Cleveland schools—which were 60 percent black and 37 percent white—and this order both exacerbated the city's shortfall and deepened white homeowner hostility against a much-needed tax levy. Teachers went on strike for thirty-six days in 1978.[44] In October 1979, Cleveland's 3,300 teachers—already one of the lowest-paid cohorts in Ohio—walked out after rejecting a 6 percent pay raise. When teachers went back to class in 1980, they received an immediate 10 percent increase and an additional 16 percent over the next two years, but the school system was left in no better financial shape. The school board's treasurer pointed out that the beleaguered school system "committed money it does not have" and was simply "trying to buy time" to keep the schools open.[45]

By far the most significant strikes in 1980 and 1981 occurred in Philadelphia. As in St. Louis, Philly's fiscal situation did not improve between 1973 and 1980, and teachers also had to contend with the aftermath of California. Furthermore, by the fall of 1980, the economy had gone into recession after Federal Reserve chairman Paul Volcker's dramatic increase of interest rates to wring inflation out of the economy caused the highest unemployment rate since the onset of World War II.[46] In addition, the Philadelphia schools were burdened with a more difficult fiscal climate after the State of Pennsylvania underwent a seven-week financial crisis in 1977. Republicans refused to pass a budget that included a tax increase, and the crisis exacerbated the Philadelphia school district's long-standing revenue problem. To stay afloat, the schools needed a major loan from Philadelphia banks. As with New York, beleaguered cities' financial difficulties increasingly positioned banks to dictate policy, and Philly banks only agreed to loans if the school system passed a balanced budget. At the end of the 1980 school year, superintendent Michael Marcase laid off 1,600 teachers to stay under budget, in addition to proposing a dozen school closings. He also raised maximum class sizes and cut prep periods.[47]

The job cuts represented the major point of contention between the Philadelphia teacher union and the board when teachers' contracts expired in the summer of 1980. With no agreement, 23,000 PFT teachers struck in September. Union negotiators acquiesced to a two-year contract with no raise in the first year and a 10 percent raise in the second (a contract similar to those recently signed by police and firefighters), but, in exchange, union president John Murray wanted all the teachers rehired. Because he had just won a hotly contested election for the PFT presidency in 1979, Murray needed to prove he could deliver at the bargaining table. More than this, however, the president—like many rank-and-file teachers—believed a capitulation would open the door to more layoffs. The board offered to rehire all of the laid-off teachers except for about 300, arguing that enrollment in Philadelphia schools would decline by 10,000 students (from 1968 to 1979, student enrollments had declined from 284,000 to 233,000, mainly a result of suburban flight). The PFT argued that the decline in enrollments allowed the district to lower class sizes below the board's proposed maximum of thirty-five students instead of laying off teachers.[48]

Five days into the strike, the board and the union reached agreement: the schools would rehire all of the laid-off teachers in exchange for contract language allowing the district to "adjust" the number of faculty if enrollment continued to decline. At this juncture, Mayor William Green III intervened to stop the agreement. Because the school board could not unilaterally raise revenues in Philadelphia, either the mayor or city council could effectively nix a municipal labor contract, and Green exercised this prerogative, arguing for language in the contract that specifically allowed the school district to "lay off" teachers if enrollments declined. "It is wrong," the mayor argued, "to saddle the taxpayers with another bill to pay for the services of personnel who are no longer needed due to declining enrollments."[49]

A mayor intervening in a municipal labor dispute had precedent. Mayor Kenneth Gibson's intervention had been instrumental in ending the long and bitter Newark strike in 1971. Mayors such as Atlanta's Maynard Jackson in 1977 pioneered the practice of replacing striking public employees during hardball labor negotiations that President Reagan would soon exercise in the 1981 PATCO strike.[50] In this case, however, Green did depart from precedent by exercising his authority as mayor to veto the agreement, knowing it would likely prolong the strike.

A Democrat who had been a U.S. representative from 1965 to 1976, Green was elected mayor in 1977. As a congressperson, he supported organized labor, but as mayor he inherited an enormous deficit from Frank Rizzo. Firmly committed to a balanced budget in the first two years of his term, he forced the city on a path to austerity similar to that in New York. During the 1980 strike, Green held firm, and after three weeks without salaries, the teachers relented, signing a contract that assented to the layoff provision in exchange for rehiring all of the previously laid-off teachers. Ominously, however, the second-year salary increases projected a budget cost of more than $90 million.[51] When school opened in 1981, a new labor conflict would emerge to make the city all but forget 1980.

A year later, much had changed. To begin with, an electoral landslide gave Reagan the presidency. The only president to have headed a union (he was president of the Screen Actors Guild from 1947 to 1952 and from 1959 to 1960), he fired and replaced 11,000 striking air traffic controllers—members of a union that had endorsed him during his campaign—on August 5, 1981. That Reagan would so decisively break a union as conservative as PATCO signified to employers—both public and private—that a new era of labor relations had dawned in the United States.[52]

Furthermore, just days before the teachers walked out, President Reagan's secretary of education Terrel Bell, citing his concern about "the widespread public perception that something is seriously remiss in our public school system," commissioned a panel to study the state of education in the United States.[53] This commission would end up publishing the report *A Nation at*

Risk, which painted a bleak picture of the American education system, in 1983. In 1981, however, the significance of the study stemmed from the fact that a growing consensus had emerged that the nation's school system was dysfunctional. In fact, public-opinion polls bear out such a trend. When the Roper Organization asked Americans, "Would you say the local schools (the ones you are familiar with) are doing an excellent, good, fair, or poor job?" in 1959, 64 percent had answered "excellent" or "good," while only 26 percent had answered "fair" or "poor." By 1972, these numbers had shifted to 50 percent and 35 percent, respectively. By 1980, only 47 percent of Americans answered "excellent" or "good," while 38 percent answered "fair" or "poor."[54]

In Philadelphia, this narrative gained additional traction in the local media: in September 1981, each of Philadelphia's largest newspapers ran a series of investigative articles about the state of the schools. The *Inquirer*'s eight-part series "The Shame of the Schools" began running before the strike started, and the *Bulletin*'s "Public Education: A Birthright in Jeopardy" ran shortly after it began. The two series painted a portrait of a failing school system, and at the core of the explanations for the failure was the unionization of teachers. The third installment in the *Inquirer* series, for instance, was titled "The Real Power in the District: The Teachers." It highlighted a poor teacher who still taught only because "firing a teacher is nearly impossible in the Philadelphia school system." The authors' assertion relied on two facts: first, that teachers needed unsatisfactory ratings for two consecutive years and then had the right to an appeal before they could be dismissed, and, second, interviews with "many principals" who said "they would never bother with such a cumbersome process except in extreme cases." Pointing to the average teacher salary, which was higher than in New York City, the piece charged that while "thousands of Philadelphia teachers are dedicated, hard-working individuals . . . collectively, their union's power has hamstrung the district."[55] The second installment of the *Bulletin*'s five-part series focused pointedly on the unionization of teachers, headlining the question, "Have Union Pacts for Teachers Gotten Out of Control?" The piece weighed the gains made by teachers since the 1960s—when they were underpaid, according to one union activist, because they "were thought of as maiden ladies, the unmarried daughters"—against the fact that teachers struck in Philadelphia more times in the 1970s than in any other city in the United States.[56]

In this milieu, in September 1981, Green told the teachers the city could not afford the negotiated 10 percent salary increase and asked them to totally forgo the raise to help close a $223 million shortfall in the school district budget. The superintendent had already used new contract language to lay off 3,500 school employees in June, an effort over which the union had threatened to strike. But Green asked for even more concessions when the new school year began. "The fact is," he argued, "that 10 years of political cave-ins, bad management, constant and continual concessions and repeated failures to heed the warning signals have brought Philadelphia's school system to the brink of complete bankruptcy." He wanted to renegotiate the union's contract and initiate another round of layoffs.[57] Green's request for concessions in a signed contract represented the public-sector equivalent of a seminal moment in private-sector labor relations two years earlier when the federal government tied its bailout of the Chrysler Corporation to a wage freeze, pension deferrals, and a giveback of paid holidays by the automaker's unionized employees. These concessions strengthened the bargaining position of the other major automakers and initiated a process of whipsawing in which employer after employer asked unionized workers to forgo wage and benefit increases or to agree to outright concessions.[58] Mayor Green, in fact, had referenced Chrysler in asking the teachers to voluntarily forgo the conditions of their contracts, asserting that "it is difficult but certainly not unusual—under extraordinary circumstances such as these—for a contract to be modified and reconsidered. . . . We saw [this] across the country in Chrysler plants. Management and labor have frequently adjusted contract provisions when it became apparent that the survival of the system, the interests of the whole and the job security of the majority were in jeopardy and that is precisely what we face today in Philadelphia."[59]

In addition to demanding concessions from teachers, Green proposed a 10 percent increase in the city's property tax. Asking for a percentage increase equal to the pay increase to be given up by teachers was an ingenious political maneuver. It allowed the mayor to argue that teachers and city residents should equally share the burden of solving the fiscal crisis (of course, losing a 10 percent salary increase represented a much larger amount of money for teachers than a 10 percent hike in the property tax represented for the overwhelming majority of Philly homeowners). Tellingly, in the era of municipal governance that had dawned after the New York City austerity program and the California tax revolt, Green did not even bother to publicly ask for more funds from the State of Pennsylvania—where the Republican governor was currently holding back $75 million over a special-education dispute—or from the federal government—where experts projected the Reagan administration's new budget to cut $17 million a year from Philly schools.[60]

Unwilling to rip up their contract—hard won in 1980—and forgo raises, the 22,000 members of the Philadelphia Federation of Teachers once again went on strike, and only about 200 teachers showed up for work the first day.[61] The strike was immediately acrimonious. Formal negotiations did not begin until early October, nearly a month into the strike. When they did begin negotiating, the teachers found themselves in a virtually impossible position as Green pressed the board for a three-year contract with a total wage freeze. On the walkout's second day, police arrested more than 200 teachers for violating a court order limiting pickets.[62] Superintendent Marcase took an aggressive stance, immediately seeking a back-to-work injunction, although judges had rarely enjoined a strike in Pennsylvania until it impinged on the completion of a full school year. Two weeks in, Marcase opened an elementary school and seven high schools (using nonstriking teachers and administrators) to teach Philadelphia's high school seniors. Striking teachers massed at the schools to prevent nonstrikers from entering, and parents and strikers screamed and cursed at each other.[63]

Efforts to open the schools also renewed long-standing racial conflicts in Philadelphia. Reports of mostly white teachers "harass[ing] black parents leading their children across PFT picket lines" led some African Americans to fiercely criticize the union. "If the Philadelphia school district was 70 percent white, we wouldn't have this," former mayoral candidate Charles Bowser reminded the City of Brotherly Love. "Nobody has a right to come into the black community who don't live in it and don't care about it." Sixteen members of the "Black community" sent an open letter to both the union and Mayor Green, asking for a "permanent solution" to the school crisis. They argued that "members of the PFT are among the best paid urban teachers in America; but they work a shorter day than most teachers in America. . . . They work in a city with high unemployment, rampant poverty, a heavy tax burden and an economy which is hemorrhaging jobs. Philadelphia and its citizens have less and less money—but the PFT screams for more and more." The letter concluded by asking teachers to "Cross the Picket Line! It will be easier than the march from Selma to Montgomery!"[64] This statement provides perhaps the best example of the wide gulf between what teachers could do and what they were asked to do. Black Philadelphians were clearly frustrated with the failure of metropolitan Philadelphia to provide quality education to city students, and the union almost certainly shared some of the blame. Still, such a statement focused that blame squarely on teachers, who were equally unable to single-handedly rectify the structural inequality between city and suburbs.

On October 7, 1981, two county judges—Edward Bradley and Harry Takiff—ordered teachers back to work (the two judges had dual jurisdiction because Takiff had unexpectedly left the city for several days early in the strike). Bradley and Takiff ruled that because the unionized teachers had already signed a contract, they could not strike under PA 195 but instead had to pursue a grievance through the state department of labor. Takiff, however, had also rendered an earlier decision that salaries included in the 1980 contract could not be enforced because the city could not pay for them. Taken together, these two decisions seemed particularly unfair to many teachers. In effect, the court told the teachers they could not strike because they had signed a contract, but that contract could not be enforced!

Once the strike threatened a full school year, the judges ordered the teachers to return to classrooms. Just as it had in 1973, an injunction instantly criminalized teachers' protest.[65] They responded with a 1,000-strong demonstration at a North Philadelphia school and marched to city hall, blocking traffic and demanding to see Green. The march attracted the support of local labor leaders: the executive leadership from the Philadelphia AFL-CIO, the Teamsters local, the National Union of Hospital and Health Care Workers local, the AFSCME local, and the post office workers local attended.[66] PFT president Murray told teachers to remain on strike, proclaiming that he would go to prison for as long as it took to restore the old contract. Judge Bradley responded by fining the union $10,000 a day.[67]

On October 15, Murray upped the ante by calling on the other major Philly unions to join teachers for a general strike on October 28, an action that the Philadelphia AFL-CIO voted to endorse.[68] Murray's call for the general strike—coeval with national television news coverage of picketing teachers screaming at "scabs" and arrested after clashing with the police— intersected with the national trajectory of the labor movement.[69] President Reagan's summary termination of 12,000 PATCO workers just two days into their August strike signified a once-unthinkable course for American labor politics. The major precedent to the PATCO walkout in the federal government had been the 1970 postal wildcat strike. President Nixon had not been able to replace the nation's striking postal employees, instead engaging in negotiations that led to a better deal for postal workers. Even PATCO, as Joseph McCartin has pointed out, staged several job actions—such as slowdowns and sick-outs—during the 1970s that the federal government met with limited retribution. Further, the job actions galvanized union solidarity, and the Federal Aviation Administration agreed to the union's first contract in 1973. Reagan's hard line against the air traffic controllers shook up the organized labor establishment.[70]

In retrospect, we know Reagan never rehired them, but in September 1981, the air traffic controllers still captured a good deal of media coverage as they sought to get their jobs back, and the broader labor movement fought back against the PATCO firing. On September 19, 1981, while the Philly strike reached its third week, between 250,000 and 500,000 union members converged on Washington, DC, for a Solidarity Day protest to demonstrate that organized labor had not lost its political potency; 10,000 of those protesters riding cars, buses, and trains to the National Mall came from the Philadelphia area.[71]

In this context, Philadelphia labor leaders publicly supported Murray's call for a general strike in late October, but rumblings from the rank and file indicated a growing rift between white-collar and blue-collar workers, revolving in part around gender distinctions. On October 27, one hundred union leaders of the Philadelphia AFL-CIO forged a "unanimous agreement" to instruct their members to walk out the next day.[72] Behind the scenes, however, many labor leaders—particularly the city's blue-collar dockworkers, Teamsters, and sanitation workers unions needed to shut down the city—doubted how many of their members would actually strike.[73] Philadelphia's teachers and blue-collar workers may have all belonged to unions, but that did not mean all blue-collar workers believed they shared the same interests. Better educated and better paid, teachers and air traffic controllers did not always arouse the sympathy of those who worked with their hands. In fact, it is possible that teachers—also white-collar workers—might have faced some of the same resentment that the PATCO workers had faced. As one union leader put it, "Our members would be giving up an $80 day for a teacher who earns $28,000? I don't know what it would be worth to us."[74] Particularly with regard to teachers, gender likely played a key role in this resentment; as research on chemical workers in New Jersey in the early 1980s has shown, male blue-collar workers (a demographic similar to Philly's blue-collar workers) often resented the predominantly female teachers who some believed made them feel unintelligent as teenagers in high school.[75]

The extent of labor support for the teachers was never determined, however, because the general strike did not take place. On the night before the strike, the PFT ended its holdout when a three-judge state appellate court panel amended the county court injunction. The higher court argued that the second year of the contract was invalid because both sides signed it under the "mistaken fact" that more revenues were forthcoming. The court ordered the teachers back to work under the terms of the first year of the contract, however—the "last valid agreement between the parties." While the new injunction did not enforce the 10 percent raise, it did immediately reinstate the 3,500 employees who had been laid off.[76] The PFT accepted the compromise and teachers went back to work. The city saved $90 million it would have cost for the raises in addition to savings from lost salaries during the fifty-day strike. Still, as teachers returned to work, the board continued to contemplate new sources of revenue to keep schools open for the full year.

Letting the Public School System Die

The major archival sources for understanding the significance of the strikes in Philly are a collection of letters to Green—especially important in 1981 because he initiated concession bargaining—and the Philadelphia newspapers. The meaning of the strikes derived from a similar context as that in St. Louis in 1979; this was the era after the tax revolt, and further tax-limitation measures made their way onto Election Day ballots, notably in Missouri and Massachusetts in November 1980. Many Philadelphians exhibited a similar producerist tax consciousness almost as soon as the three-week 1980 strike began. Maria Louisa Pepper, for instance, wrote to the *Bulletin* on September 21 that she had been victimized: "If Mr. Murray and the teachers paid their tuition at college and their professors went on strike, they would be the first to yell and demand a rebate. I feel that my school tax was my children's tuition, so I demand my rebate."[77]

The *Inquirer*'s assessment of the 1980 strike also placed the taxpayers at the center of the controversy, asserting that "Mayor Green deserves the support of every Philadelphia taxpayer, if not the teachers union, in holding the line for a fiscally responsible contract." The newspaper's editor was Edwin Guthman, a liberal Democrat who had served as Senator Robert F. Kennedy's press secretary. After working for Kennedy, Guthman moved on to the chief editorship of the *Los Angeles Times*, coming to the *Inquirer* in 1977. The editorial page's assertion, then, of the imperative of a "financially responsible contract" did not merely represent the opinion of a conservative antiunion ideologue. The newspaper's view—that "overly generous contract settlements and unbalanced budgets are what have put the school system in its present financial chaos"—had become almost common sense in American urban politics after the New York City fiscal crisis. Tellingly, this narrative lacked anything at all about the American metropolitan structure or the special expenditures urban school systems faced in the 1970s.[78] The same editorial page implored teachers, several days later, to "face reality" on layoffs, asking, "If public school enrollment continues to decline . . . should the taxpayers be required to keep on paying teachers to teach non-existent pupils? That's featherbedding and can't be defended responsibly."[79]

Philadelphia's second major newspaper—the *Bulletin*—also focused its assessment of the strike on the city's taxpayers, arguing in the conflict's second week that Green was "correct in insisting that the schools—and more fundamentally, the city's taxpayers—simply can't afford to carry teachers who have no students to teach. We urge Philadelphia's public school parents, as parents *and* taxpayers, to stand with Bill Green on the layoff issue."[80] As the strike concluded, the newspaper lauded the mayor for holding out in the interests of Philadelphia's productive citizens "for the only deal that should be acceptable to the people who pay the taxes."[81]

As public discussions continued to focus on taxpayers, questions emerged regarding how teachers spent their time and whether their pay was justified. In other words, to use a market metaphor about production, did the taxpayers get their money's worth? Acel Moore, in an opinion piece for the *Inquirer*, admitted that "teachers deserve" a good salary, health insurance, and summers off. However, he asked, "What assurance do parents have that teachers are using their prep time to prepare for their lessons? Do they leave early or go shopping or do other things?"[82] William McGuire, from suburban Ambler, proclaimed that teachers rode a "gravy train": "Average pay $23,000 for nine months work. Efficient or non-efficient workers all the same as long as they pay union dues. Oodles and oodles of fringe benefits. Wake up, America!"[83] Betsy Dubb, from Philadelphia, argued that in spite of the demands of teaching, schools required limited instructional time: "[If] teaching more than four hours a day is [too much], perhaps [teachers] should use their prep times to seek employment in another field."[84] Philadelphian Cecile Wright believed teachers should "put in an extra 50 days a year giving remedial classes to the children they have failed to educate properly during the normal school year."[85]

Teachers lashed out at the critics who purported they were not worth their tax dollars, arguing that they were not parasites, but that Philadelphians did not understand the demands of their work. Johnnie Wiedmann responded directly to Betsy Dubb by pointing out the difficulty of the job. "Teachers," he argued, "are called on to be alert and attentive every moment of their working day. . . . In addition to being knowledgeable in their field, teachers know the name of every student they've had as well as the 150 they have now. . . . Teachers deal daily with a system insensitive to their needs, insulting to their intelligence, and insistently ignorant of their humanity."[86] Finally, Myra Kane pointed directly to the large class sizes she believed would result from the 1980 layoffs: "Saving money by firing teachers is very shortsighted. . . . If the school district kept its teachers, who are the key ingredient in education, and because of lower enrollments, reduced the class sizes, Philadelphia would have a fantastic educational system."[87]

In 1981, the gap between "taxpayers" and teachers was even wider, particularly because Green tied the tax hike to teachers' forgoing raises. The

Greater Philadelphia Area Chamber of Commerce, for instance, wanted to ensure that no new business taxes ended up as part of a solution to the crisis. In 1973, the same body had publicly called for more taxes on Philadelphia businesses because of the public importance of education. Now, however, the chamber asserted, in a letter from chairman-elect Harold Sorgenti to Green, that "this community has reached its limit. The law of diminishing returns, so far as taxes are concerned, is now firmly entrenched in this city's economy." Sorgenti concluded that, "based on declining enrollment, significant cuts *can* be made."[88]

The *Inquirer*, as it had in 1980, rallied to Green's side and argued for a long-term solution to the deficit. On September 3, for example, the newspaper argued that the only "possible source" to solve Philly's long-term problems was its "taxpayers." It pointed to the taxpayer's average obligation—around $4,000 per student enrolled in the system—and argued that all they received for the money was "an unmanaged, unprofessional out-of-control, tragic burlesque of what they deserve and have every right to demand. Was it irresponsible for the men and women who represent Philadelphia's taxpayers to negotiate the series of teachers' contracts that created those staggering deficit figures? Without the slimmest doubt, yes." Mayor Green represented the city's last hope, and "for him to step back from it would be to behave as irresponsibly as did those who created the crisis."[89]

Disgruntled taxpayers wrote to both Philly newspapers to side with the Democratic mayor. Philadelphian Stephen Lawrence characterized himself as an "overtaxed citizen" who "resent[s] an additional increase in my taxes." Showing how well Green's effort to link his breach of the teachers' contract to the tax increase had worked, Lawrence admitted he would be willing to pay more taxes if the teachers would also forgo the raise and accept layoffs. He argued that PFT president Murray's refusal to relinquish the raises "shows absolutely no concern" for the city. Lawrence further connected the strike to the wider labor movement, arguing that teachers' demands undermined the position of unions more generally: "Labor leaders like Mr. Murray are one of the reasons that the labor movement in America is losing credibility and, it would seem, strength." Donna Greenberg made an explicit connection to the productive population that appeared to be leaving Philadelphia. She argued that the "taxpayers of this city" represent a "rapidly shrinking population. . . . Eventually there will be almost no one left to pay for the schools. . . . I recommend that we put our collective foot down. The time is ripe: public sympathy is turning against unions. We could use some of President Reagan's hang-tough philosophy here in Philadelphia."[90]

Homeowner C. J. Swartz invoked California when he compared his property taxes to his son's, who lived in the Golden State in the post–Prop 13 era. Sent in an envelope featuring a sticker that read "Cut Taxes," Swartz's letter pointed out that his yearly property tax had increased dramatically from 1974 to 1981 (he asserted that it had almost doubled). He urged Green to "resist this assault on the people of Phila. by the teacher's union. . . . My son lives in California. His house has more than 4 times the sales valuation as mine," but he paid about the same property tax.[91] Other commentators, like O. M. Monroe, criticized both teachers and the mayor. He called Green's solution "disgusting and most unfair to the average taxpayer, who is now over-burdened with taxes and higher utility costs, to be hit by yet another increase." Monroe believed school financing should more closely resemble a pay for services structure, as "most people who have purchased real estate in the last 20 or 25 years have no children in the schools."[92]

In a lengthy letter to Green, Carol Aff, writing from white working-class Northeast Philadelphia, turned what she believed were the assumptions about education in the city on their head. "Everyone is so concerned about paying lip service to the 'children' in this city," she pointed out, "that it seems repugnant to mention that there are elements in our city who are more important than the children." In thinly coded racial language, Aff starkly separated the interests of those she believed produced from those whom she assumed did not:

> Those of us who pay taxes <u>are</u> more important! We're more important than the kids who don't want to go to school and don't learn anything anyway. We're more important than the 70% of the parents who are on welfare and provide nothing to further their children's education not even the discipline that is lacking and which bears a lot of responsibility for making the system the mess that it is. We foot the bills for too much waste and too many unnecessary programs and by and large we either don't have children or wouldn't send them to public schools if we did.

Aff's use of the term 70% could hardly have been accidental, as anyone paying attention to the school system knew it was 70 percent black. She suggested that Green "cut out everything, and I mean everything that is not required by law. . . . No one would like you for it, except for the taxpayer who would smile to themselves and applaud your efforts." Connecting education to the market, she admonished the mayor to "get educators off the school board. Education is a business and people with business backgrounds should be running the show."[93]

Beyond rebuking the teachers and tax increases, some critics advocated direct action in the mold of Prop 13 and Reagan's stand against PATCO.

Two weeks into the 1981 strike, for example, Philadelphian Penny Brodie proclaimed to readers of the *Inquirer* that "it's time to withhold your school taxes; it's time we had enough with the teachers' union; it's time for the school board officials to say to the teachers, 'Show up tomorrow or you're fired.'"[94] Others directly threatened Green's political career if he followed through with the tax hike. An "irate taxpayer" from Northeastern Philadelphia, for example, asserted that "if a tax increase is put into effect, we the Tax Payers who put you into office, will vote you out."[95]

The occasional taxpayer did assert their duty to pay higher taxes. Philadelphians William and Gloria Powers wrote to the *Inquirer* on October 4, 1981, that "if it takes higher taxes to support [the school] system and insure its viability then we are prepared to pay them. To do otherwise is irresponsible."[96] Another citizen, from Bryn Mawr, went even further, sending Mayor Green a check for the not insubstantial sum of $236 (one-millionth of the estimated budget shortfall) on October 20. "If citizens would think only of what this [strike] is doing to pupils . . . and think of ridding 'our' City of this most sad, sad situation I hope most all who are able would give freely and promptly."[97] Such views, however, represented clear exceptions.

In fact, most of the voices defending the strike were teachers, and, in contrast to 1980, when teachers affirmed their professional value, now they predominantly parried critics by emphasizing the city's reneging on a signed contract. Murray, for example, reminded Philadelphians that "our members voted overwhelmingly, in good faith, for this contract knowing full well that the wage and benefit package was to be deferred for one year. . . . In plain language, the mayor wants to change the rules in the middle of the game to the mutual detriment of school employees and school children." A striking teacher pointedly criticized the *Inquirer*'s support for Green, asking, "Why after having been on strike for 21 days last September to settle a contract that got me no raise last year, I should not be looking forward to and fully expecting this year the raise that I have already suffered for?" Jane Schreiber also criticized the *Inquirer*'s view of the strike by comparing its coverage to that of PATCO: "On the one hand, you criticize the air traffic controllers' strike, because the controllers failed to honor signed agreements with the federal government. . . . On the other hand, you condone the city government's failure to honor a signed agreement with the Philadelphia Federation of Teachers."[98]

Other teachers located the struggle in the larger trajectory of American labor relations. Julius Rosenberg argued that if the board "is permitted to break this contract it will have set a precedent which will threaten the entire process of collective bargaining all over the country." Similarly, Jane

Henderson asserted that "if the words *binding* and *contract* still have any meaning, then this request [to forgo the teachers' raise] is utterly unreasonable. Why waste time and energy and school days negotiating a meaningless contract? Backing down now would only mean more of the same, not only for the teachers, but also for workers in other unions."[99]

Not all working people, however, believed the teachers' fight had much to do with them, and many believed the line dividing the teachers from other workers centered on gendered notions of productivity. Betsy Dubb, for instance, discursively separated teachers from other workers when she argued that "the average American puts in an eight-hour work day with an hour for lunch. He enjoys two weeks' vacation a year. His pay and benefits lag far behind those of [teachers]." By employing the masculine pronoun, Dubb signaled that the normative worker was a man, as opposed to those in the teaching profession, consisting mostly of women. In a similar vein, Steven Klotz pointed out that "most people I know work far longer hours than do teachers, not because they are paid overtime, but because there are not enough hours between nine and five to accomplish their tasks in a satisfactory fashion. Pride in one's work—a characteristic most teachers claim to possess—commands one to labor until the job is done right and done well."[100]

Others more starkly emphasized the space between teachers and other workers. B. Defere believed teachers "should understand that the handsome wages and fine fringe benefits they receive come right out of the hides of those who do not get them—senior citizens and low-paid workers." Mary Rogrove, a teacher from Kempton (a town seventy miles from Philadelphia), excoriated the PFT for "misunderstanding . . . the difficult position of the rest of the taxpayers who support them." She believed that "we should sympathize with these other taxpayers who are suffering more from our floundering economy because their salaries are not guaranteed as ours are." Also defining the normative worker as male and blue collar (she used the example of a carpenter), she believed that he "now suffers the anxiety of building his trade when money is tight, and . . . working extra hours for less money when market demand for his skill drops off. He has the same bills we have, but none of the job assurances guaranteed to us in the teaching profession."[101] By early 1981, then, the support of other workers—perhaps even many unionized blue-collar workers—for striking teachers was far from certain. In this context, it is a highly interesting counterfactual to wonder what would have happened had the court not made a face-saving ruling before the general strike. Indeed, the PFT and the Philadelphia AFL-CIO might have been as worried as anyone else about the possibility.

In addition to widening the gap between teachers, other workers, and tax-payers, the strike—equally importantly—opened up a space in which commentators began to imagine alternatives to the public school system. Indeed, the lengthy, intractable financial problems coupled with the print exposés about the failure of the education system led many to question whether the schools should ever reopen. Dan Rottenberg's op-ed in the *Inquirer*, for example, compared the Philadelphia schools to the seventeenth-century Dutch tulip trade and Major League Baseball (which had experienced a players' strike in the summer of 1981). All three, he argued, relied on "maintaining a public illusion that your commodity is worth taking seriously." Although education, according to Rottenberg, did possess "intrinsic" value, the school administrators and teachers squabbled under the fallacious assumption that students would return no matter what. "Yet each successive strike," he concluded, "adds fuel to the notion that Philadelphia's public school system is nothing more than a gargantuan tulip collection. This summer many Philadelphians discovered that they can get along just fine without baseball, this fall many Philadelphians are discovering that they can get along just fine without the public schools." Similarly, Robert Kay, from suburban Paoli, pointed out that "in our anxiety to get the schools going again it's a pity that few seem to be asking the really important questions." He asked the readers of the *Inquirer*, "Are the schools, in their present form, really worth salvaging?" Philadelphian Charles Hewins's answer: "If we can't beat the system, let's learn how to establish a new one."[102]

But if the public school system had failed, then what was to be done? The alternatives discussed in Philadelphia all revolved around market-based systems: new schools with nonunionized teachers, private schools as competition, tuition or head taxes, or, to bring it all together, a system of "vouchers" to instill competition by allowing students to attend whatever school their parents chose.

Edna Williams wanted Mayor Green to close the schools and reopen them with nonunion teachers. Writing from Northeastern Philadelphia, Williams believed that "between the union demands and welfare recipients, the nonunion workers and home owners are the victims of these greedy, power and money hungry people." She also wanted Green to know that unions "have bullied us long enough and it is up to our leaders . . . to say 'No' to their demands. . . . Many think the School Dist. should declare bankruptcy, fire these striking teachers, and set up a whole new school system with non-Union teachers. Many of the older citizens and homeowners live in fear of losing their homes, when taxes are raised to meet the demands of these selfish people."[103]

Patrick Armstrong, also from Northeastern Philly, wanted to shift the burden of paying for schools from homeowners to parents whose children attended public schools. He also had strong opinions about who produced value for the city and who did not. "I suspect the majority of the children attending public schools are not, in fact, children of property owners," he argued. "The parents of children who attend school should pay tuition of some type thereby having the burden of the cost directed toward those who benefit. I would urge keeping the property taxes of tax paying citizens of Philadelphia at a reasonable level so that the few responsible citizens of the city will not leave."[104]

The private school system represented an obvious alternative to many critics of Philly schools. Catholic schools had emerged as an alternative to compulsory public schools as early as the nineteenth century, particularly in cities with large working-class European immigrant populations like Pittsburgh, Milwaukee, and Philadelphia. Private schools also emerged as alternatives to integrated schools in the South during the white establishment's massive resistance to civil rights. By the 1960s and '70s, private schools increasingly served as the refuge of white ethnics from court-ordered efforts to desegregate urban public schools. Now, critics offered them as an alternative to what they viewed as failing schools. In late September, the *Bulletin*, for instance, published an op-ed from the "parents of three children who up until this year attended public schools in Philadelphia." Russell and Rosalind Jackson "believe[d] in the concept and value of public education—properly administered. . . . Under present circumstances, private school offers the best short term option. For the longer term, we will, as many others already have, get out of this town as fast as we can and find a place where public education has not become the rotten plum of decades of selfish, unconcerned politicians and school personnel."[105] Philadelphian R. Tucci advocated charging tuition to everyone: "We, who send our children to private or Catholic schools, pay a tuition plus our school taxes. Why should we pay school taxes if we have to pay tuition? Let them pay." As the strike dragged into its seventh week, John Karpowicz had bigger ambitions. He argued that "the people of Philadelphia should send their children to private schools and organize a tax revolt."[106]

Such drastic overhauls, however, were not entirely realistic given the long history of the state providing education in the United States and education rights enshrined in most state constitutions. Proposals for school vouchers, however, emerged as the most prominent systematic alternative during the 1981 Philadelphia strike. Voucher systems represent a hybrid between state financing for education and parental choice in schools. Although there are many possible forms, in general, a voucher system gives individuals a specific amount of money that can be used to attend private or public schools. The logic is market based—parents supposedly can best discern the appropriate school for their children, and competition for their state-funneled dollars will in turn create a better overall educational "product."

First proposed by economist Milton Friedman in a 1955 essay, "The Role of Government in Education," Southern segregationists sought vouchers in order to fight court-ordered integration (called, at the time, "tuition grants," which white students could use at private schools). Vouchers also gained prominence by the 1960s and 1970s as an experimental idea among free-market fundamentalists as well as some liberals, like Harvard education professor Christopher Jencks, to desegregate schools. In Nixon's Office of Economic Opportunity (OEO), Donald Rumsfeld and others sought to use federal funding to promote experimental voucher systems. The OEO pushed unsuccessfully, for example, to get local school boards in New Hampshire to develop viable voucher programs.[107] Neoliberals did not give up, however: *National Review* publisher and prominent Ronald Reagan backer William Rusher believed that the political program of the "new majority" third party he outlined in 1975 should include "an unqualified school voucher system in each state" in order to "allow parents (as opposed to teachers and the public education bureaucracy) the maximum opportunity to select their children's education milieu." Not only would this system provide choice, but it would also allow Americans to "reconsider" the "whole underlying assumption of our American educational system—that everyone will benefit from as much scientific or liberal arts schooling as possible." It would spare both students and taxpayers "where it is obvious that certain students are not obtaining a serious education."[108]

Only in 1980, however, did vouchers truly enter the national political mainstream, through Milton Friedman's ten-part Public Broadcasting Service (PBS) series *Free to Choose*. After winning the Nobel Prize in Economics in 1976 for his work on inflation, Friedman enjoyed substantial intellectual currency, which he cashed in for the widely popular series. Conceived by a local Pennsylvania PBS station manager, the series received no actual funding from PBS; instead, financing came from the Sarah Scaife Foundation (a conservative think tank named after the mother of Richard Mellon Scaife), Getty Oil, and *Reader's Digest* (which also ran advertisements to promote the series). In January 1980, around the time a book of the same title reached the *New York Times* best-seller list, the hour-long segments began to air weekly.[109] The first half of each episode featured a different aspect of society that would be dramatically improved, according to Friedman, through less restriction of market forces. Then, the second thirty minutes included a discussion with relevant experts. Episode 8, "Who Protects the Worker?," argued, for instance, that unions actually harmed most workers by restricting competition. In the roundtable, Friedman debated his thesis with the assistant secretary of labor, union leaders, and business executives.

Episode 6—"What's Wrong with Our Schools?"—focused on the flaws with "government schools." The segment began with footage from an inner-city Boston school, highlighting the school's police officers and metal detectors, as Friedman's avuncular voice-over asserted that "nobody is happy with this kind of education. The taxpayers surely aren't. This isn't cheap education. After all, those uniformed policemen, those metal detectors have to be paid for." He argued that wealthy Americans could opt out of subpar schools by relocating or simply sending their children to private schools. "Increasingly," Friedman continued, "schools have come under the control of centralized administration, professional educators deciding what shall be taught, who shall do the teaching, and even what children shall go to what school. The people who lose most from this system are the poor and the disadvantaged in the large cities. They are simply stuck. They have no alternative."[110]

In the economist's view, choice would promote better education because schools would compete for students. Specifically, Friedman advocated vouchers, highlighting a small-scale example in which the reform had improved education. He then used the example of a school in England to show how local "bureaucracies"—particularly unions—blocked change: "For four years, there have been efforts here to introduce an experiment in greater parental choice. Parents would be given vouchers covering the cost of schooling. They could use the voucher to send their child to any school of their choice. I have long believed that children, teachers, all of us, would benefit from a voucher system. But the headmaster here, who happens also to be secretary of the local teacher's union, has very different views about introducing vouchers."

Friedman here used the United Kingdom—which had a reputation for stronger union power than the United States, particularly after the 1978–79 Winter of Discontent—as a cautionary tale for how American teacher unions would try to stifle reform. The first half of the episode ended with Friedman's conclusion that "what we need to do is to enable parents, by vouchers or other means, to have more say about the school which their child goes to, a public school or a private school, whichever meets the need of the child best. . . . Market competition is the surest way to improve the quality and promote innovation in education as in every other field."[111]

In the panel discussion that followed, Friedman joined a spirited debate with different education experts, including voucher proponent John Coons (coauthor of a policy piece widely cited in the 1970s, 1980s, and 1990s—*Education by Choice: The Case for Family Control*[112]) and Albert Shanker. The latter, unsurprisingly, defended the public school system from the market forces Friedman championed. "You cannot have free competition," Shanker argued, "where one group of schools must accept every single student who comes along, no matter what his physical or emotional handicaps or other problems; whereas the very essence of a private school and your voucher school is that they're going to be able to keep out the[se] students." Shanker also worried that vouchers would destroy the public obligation to provide equal education to all students because a permanent group of "problem students" would be ghettoized into the public schools forced to keep them. Friedman admitted that "there are a small minority of people who are problems," but he nevertheless wondered:

> Is it desirable to impose a straightjacket on a hundred percent of the people, or ninety percent of the people, in order to provide special assistance or special help to four or five or ten percent of the people? Not at all. I think that there's a big difference between two kinds of systems. One kind of system in which the great bulk of parents have effective freedom to choose the kind of schools their children go to, whether it's the lower or the higher level. And there are programs and provisions for a small minority. That's one kind of a system. That isn't what we have now.

Shanker countered in the segment's concluding point by comparing the voucher system to "a hospital throwing out all the sick patients and keeping the healthy ones."[113]

Shanker may have had the last word in *Free to Choose*, but if the public's opinion about the two Philadelphia teacher strikes was any indication, Friedman's view had grown in political legitimacy. During the 1980 strike, *Inquirer* columnist Dan Rottenberg pointed specifically to *Free to Choose* and to Friedman's voucher plan, weighing the economist's "optimistic idealism" against Shanker's "gloomy realism." He hoped that the Philly strike would "force us to . . . come up with some creative long-range alternatives—perhaps like the voucher scheme Friedman proposes."[114]

In the fall of 1981, louder voices advocated a voucher plan to change the way the Philadelphia schools operated. The *Bulletin* led the charge. After it became apparent that Green's austerity measures would provoke a strike, the *Bulletin* did more than just support his stance. Comparing the situation to private-sector labor relations, the newspaper believed a voucher program could empower the board to break the strike: "If the teachers decide to strike, the city conceivably could offer a voucher system which would pay parents to find alternate schools or means to educate their children."[115]

Although the basic view of the *Inquirer*'s editorial page did not change much between 1980 and 1981—the newspaper never directly advocated vouchers—it printed several op-eds by Neil Peirce advocating a voucher overhaul of the American school system. Published during the 1981 strike, one piece specifically linked teacher strikes with taxes and implored the reader to "look across the political and educational wreckage of many major city school systems this fall—militant unionism, intransigent bureaucracies, forced busing, taxpayer revolts, skyrocketing costs, and slumping student achievement scores." In this context, he asserted that in Philly, "for the sake of the kids, if not the taxpayer, hasn't the time finally arrived to break the effective monopoly of the public school system? Why not experiment with a system of vouchers to permit parents to shop for the school they believe would provide their child the best education?"[116]

Philadelphia parent Geoff Steinberg was "glad to read" Peirce's case for vouchers. "How wonderful and convenient it would be if the public school one block from our home were an adequate educational institution," he pointed out. "But it is not, and we are not willing to gamble and assume that the schools will improve." D. B. Stad's support for vouchers emerged directly from the strike: "Children need schools they can depend on and teachers who care about them and maintain a professional attitude toward their work. Let's get on with concrete plans for a voucher system and financial support for alternative schools as soon as possible." Gerald Palladino believed vouchers represented "the only way out." He argued that "the school system [sh]ould have to compete. Those not measuring up to the standards of the people desiring education will fail, as they should."[117]

By the end of the 1981 strike, then, public discussion of teacher unions and the education system were vastly different than they had been in 1970 when the Pennsylvania legislature passed its new public-sector labor law. In 1970, policy makers had tried to improve the way local governments bargained with their workforces and hoped this would create a better-functioning education system. In 1972 and 1973, many Philadelphians wanted to know how to solve the structural problem of financing the city's public schools, and this debate initially included a far-ranging critique of metropolitan political economy; by the end of 1981, in contrast, many Philadelphians instead sought systematic ways to recalibrate the public obligation of educating the city's students around market behavior.

Conclusion

Not everyone in Philadelphia believed voucher schemes would dramatically improve the education system; as might be expected, teachers expressed particular skepticism. Elliott Rubin, for example, characterized vouchers as "an effort in America to destroy public education as we know it because the expense in dollars seems to far exceed the return."[118] Rubin's framing of the issue in terms of expenses and returns represented an accurate appraisal of how many Americans increasingly assessed education by the beginning of the 1980s. A decade of economic difficulty, inflation, skepticism toward the efficacy of the state, mounting taxpayer consciousness, and the rising influence of neoliberal economics represented by the ascent of Friedman shifted the prism through which many Americans understood both economic opportunity and social equality. At the end of the 1960s, many Americans still believed that teacher unions ensured decent salaries and working conditions for their members while improving the education system—even if they shouldn't strike. Many Americans also believed that state-supported education represented more than just a set of economic inputs and outputs. The robust debates over school financing that we saw in Philadelphia and St. Louis in 1972–73 highlight the purchase of these assumptions. By the end of the strikes examined in this chapter, however, it is clear that many Philadelphians and St. Louisans viewed unionized teachers and the students they taught as unproductive and harmful to the city's political economic structure; a deluge of commentators in both places focused on the results the taxpayers were (not) getting for their money. In President Ronald Reagan's inaugural address in January 1981, he famously argued that "in this present crisis" of a stagnant economy, rampant inflation, and staggering energy costs, "government is not the solution to our problem; government is the problem."[119] By the end of the strike in 1981, it was clear that many Philadelphians could add the intractable difficulties of the city school system to this assessment: government was not the solution to the problem, government and its unionized workforce *had become* the problem.

TEACHER UNIONS AND THE AMERICAN POLITICAL IMAGINATION

The 1981 Philadelphia strike was the last really long urban teacher strike in the United States. Indeed, the number of significant strikes declined dramatically after the period of this study. In part, this decline resulted from the Federal Reserve's effort (under Volcker's stewardship) to stamp out inflation. With lower inflation and then a rebounding economy in the 1980s, neither municipalities nor public employees had to fight quite as urgently as they had under the toxic combination of economic stagnation and inflation. Furthermore, as we have seen, political common sense for many Americans about the role of the state and labor unions changed, which limited public support for strikes by unionized urban teachers by the late 1970s. Urban teacher strikes still occurred in the 1980s: Chicago teachers, for example, walked out for two weeks in 1983, ten days in 1984, and in 1987—the longest teacher strike in the city's history—for nineteen days.[1] Not surprisingly, in the midst of the 1987 strike, a drugstore executive who had founded the Civic Committee of the Commercial Club of Chicago called for reorganizing the education system along market lines by instilling competition among schools for students and merit pay for teachers.[2] Still, whereas in 1975 alone there were 241 teacher strikes, by the fall of 2005, as *Education Week* reported, teacher strikes were "limited to a few small districts." According to a right-wing think tank, from 2000 to 2007, there were only 137 total teacher strikes in the United States. More than half of those were in Pennsylvania, even though strikes were down in the Keystone State following a revision to PA 195 in 1992 making it more difficult for teachers to walk out.[3]

The decline in strikes, however, did not mean that teacher unions were broken, nor did it reduce the significant position teacher unions held in education and American politics. On the contrary, the AFT currently has about 1.6 million members, and the NEA, at 3 million, represents the single largest labor union in the United States. Indeed, a search for the terms *teacher strike* and *teacher union* in Google Books Ngram Viewer underscores this phenomenon. In the 1900s and 1910s, about 0.00000010 percent of books mentioned the term *teacher strike*. From World War I until the mid-1930s, there was virtually no mention of strikes. There was a spike in the term in the mid-1940s—almost certainly as a consequence of the post–World War II strike wave. After a drop-off in the 1950s, *teacher strikes* reached an all-time high, with more than five times as much ink as in the post–WWII strike wave, in the early 1970s. After that, the percentage of books discussing *teacher strike* declined dramatically, and, by 2007, it was back to the level around the time of World War II. The interest in *teacher union*, however, only increased—from a number lower than *teacher strike* until the mid-1970s—and peaked in 1995 at ten times more than *teacher strike*.[4]

One reason for this increased attention to teacher unions certainly stemmed from their emergence as a large and permanently important institution. Significantly, however, teacher unions have also clearly captured the imagination of those interested in changing American education. Even as teacher strikes declined in importance, unions continued to inspire exposés similar to the one Robert Braun wrote about the AFT back in 1972. Books in the past two decades with titles like *The Teacher Unions: How the NEA and AFT Sabotage Reform and Hold Students, Parents, Teachers, and Taxpayers Hostage to Bureaucracy* by Myron Lieberman, George W. Bush secretary of education Rod Paige's *The War against Hope: How Teachers' Unions Hurt Children, Hinder Teachers, and Endanger Public Education*, and Terry Moe's *Special Interest: Teachers Unions and America's Public Schools* leave little mystery about their arguments regarding the relationship of teacher unions and the state's provision of education in the United States.[5]

Teachers have clearly also captured the imagination of those holding or interested in holding political office, largely because, as the developments of this book have shown, criticizing them represents a form of political currency. For much of the Right, in particular, teacher unions represent the bête noire of the hated public-sector labor movement. In Wisconsin's dramatic conflict in 2011, for example, Republican governor Scott Walker successfully stripped most public-sector workers of meaningful collective-bargaining rights, arguing—using language emergent in this study—that unionized public employees like teachers were the "haves" while private-sector workers represented the "have-nots." That struggle—which he has since compared to taking on the depraved terrorist group that calls itself the Islamic State—represents Walker's signal political achievement and catapulted him into national political stardom.[6]

Similarly, onetime presidential contender Chris Christie—a Republican governor in a traditionally blue state, New Jersey—has used the perception of his tough stance against public-sector unions, especially teachers, to forge his own political reputation. Indeed, in Christie's formative national political moment—a keynote speech at the 2012 Republican National Convention—he argued that a major difference between Republicans and Democrats is that the latter "believe in pitting unions against teachers, educators against parents, lobbyists against children. They believe in teachers' unions. We believe in teachers."[7] Later, a reporter for CNN baited Christie—struggling for attention in a crowded primary field—with the question, "At the national level, who deserves a punch in the face?" The governor's instinctual reaction: the "national teachers union" (referring to the AFT): "They're not for education for our children. They're for greater membership, greater benefits, greater pay for their members. And they are the single most destructive force in public education in America. I've been saying that since 2009, and I've got the scars to prove it."[8] These bellicose discourses from the Right are even more affronting considering that the teaching profession has become consistently feminized each year since the events in this study, and women now represent almost 80 percent of the nation's teaching force.[9]

Republicans do not have a monopoly on criticizing teacher unions. Indeed, although the AFT and NEA spend millions each election cycle supporting Democrats, there are plenty of liberals who blame unionized teachers for stifling market-based efforts to reform the American education system. District of Columbia schools chancellor Michelle Rhee (2006–10)—appointed by Democratic mayor Adrian Fenty—rose to national fame through her well-publicized efforts to dismantle the structure of educator unions in the schools of the nation's capital.[10] The prize-winning film from 2010—*Waiting for "Superman"*—blames teacher unions in large part for the failing American education system while arguing that charter schools—which can circumvent collective-bargaining agreements—represent its salvation. The film, Davis Guggenheim's follow-up to his highly sympathetic documentary of Democrat Al Gore's effort to show the urgency of climate change, *An Inconvenient Truth* (2006), was far from a Republican mouthpiece. During the nation's last prominent teacher strike—by Chicago teachers in September 2012—Democratic mayor (and former White House chief of staff in the Obama administration) Rahm Emmanuel publicly clashed with the Chicago Teachers Union over school closings, teacher evaluations, and the lack of basic necessities for the city's students.

The highly politicized role of teacher unions since the 1970s underscores the inextricability of teacher unions and expectations about what the state can accomplish since the New Deal. *Teacher Strike!* has shown that in the years after World War II, a central assumption of American politics was that interventionist state action, aided by robust labor unions, was necessary to advance social equality and economic justice. The economic and political crises of the 1970s undermined this notion, but only because of very specific developments. Indeed, the public-sector labor movement in American cities came of age at the exact moment that, first, African American activists organized to rectify the abject inequality that New Deal liberalism helped to institutionalize, and second, cities faced both declining tax revenues and taxpayer resistance. Contentious teacher strikes exacerbated an already overwhelming sense of crisis in the decade and interjected anxieties about the direction of the country directly into the everyday lives of many Americans in these cities and in the suburbs surrounding them. Indeed, the insolubility of teacher strikes in the 1970s resulted from fundamental problems that labor liberalism could not adequately solve: whether public-sector workers should have full union rights or cede their most important bargaining leverage to the state and the spatial and racial inequalities that emerged from the structure of American federalism. These failures represented part of what led many Americans who had comprised the powerful labor-liberal coalition to reassess the axis on which they viewed the country's basic political divisions.

The political networks from the right emerging in the 1970s—increasingly arguing for a return to the classical liberalism of the nineteenth century—tapped into long-standing racial conflict, cultural assumptions about "productive" citizenry, anxiety about shifting gender roles, and the beliefs of much of the white working and middle class that the state victimized them during a tough economic climate. This "neoliberalism"—arguing that those who worked the hardest and produced the most deserved the most rewards—increasingly shifted the expectations Americans had about opportunity and security: only individual competition in the marketplace—not collective organization or social policy—could provide it.

This obsessive focus on markets and privatization has had a number of detrimental effects on the American education system. As education historian Diane Ravitch has powerfully shown, No Child Left Behind's unrealistic premise that every single student—no matter what legitimate structural impediments might hold them back—should be proficient in math and reading has furthered calls for even more market-based reforms like charter schools. The Obama administration's signature education program—Race to the Top—was also built around competition: in its first round of grants, it provided funding only to a limited number of states most willing to recalibrate their education policies around such things as charter schools and teacher-accountability programs. Further, the discourse of failing schools has allowed corporate "reformers"—some of which, like the Bill and Melinda Gates Foundation, appear to be making good-faith efforts to improve education based on market principles, and others, like the American Legislative Exchange Council (ALEC), that simply seek profit opportunities for its members by privatizing various public goods—much more influence over public education policy.[11]

None of this has done anything to stabilize education for the nation's neediest students. In Philadelphia, for example, the fiscal outlook has not improved since the 1970s, in part because charter schools in the district—some of which are managed by private corporations—have cut into district enrollments, siphoning off resources and exacerbating an already tenuous budget situation.[12] In fact, a drastic shortfall leading into the 2014 school year forced the city to institute a steep two-dollar-a-pack cigarette tax that disproportionately affects working people in Philly who cannot go outside the city for smokes. Further, in October 2014, the state-appointed guardian of the schools—the Philadelphia School Reform Commission (SRC)—unilaterally canceled teachers' contracts, ordering them to pay more of their health insurance premiums in order to close the budget gap. In forcing the givebacks, the chair of the SRC asserted that "the time has come for [teachers] to share in the sacrifices everyone else has made."[13] If that statement's reminiscence to Mayor Green III's call for austerity in 1981 isn't eerie enough, one need only note that SRC chair's name: William Green IV.

Indeed, there are very good reasons to be skeptical of the drift toward an America in which market competition seems to undergird most social and economic policy. Even beyond what it seems to be doing to public education, long historical examinations of capitalism, such as that of economist Thomas Piketty, clearly show that when market competition is left to its own devices, inequality tends to increase. Indeed, Piketty's *Capital in the Twenty-First Century* uses just about every imaginable piece of economic data to show just how dramatically unfettered capitalism deepens the advantages that the wealthy—or, in more specific terms, those employing capital—have over everyone else.[14]

In this context—and in a context in which political debate has become so vicious that one recent presidential candidate hopes for assaults on teachers and another compares them to terrorists who brazenly execute innocent people, and both do so with no negative consequences—what role is there for those who envision a world that prioritizes satisfying jobs, meaningful education, and truly rewarding lives for more than just a lucky few? As we have seen, teachers in this study dealt with a general lack of respect, gender discrimination, unfair labor laws, a disproportionate amount of the blame for metropolitan inequality not even mostly of their own making, and blame for fiscal crisis not of their own fault at all. My goal here, however, is not to assert that the teachers in this study were always right for striking. A reasonable assessment, in my view, is that, while virtually every strike described in this book was understandable, not every one was necessarily defensible, and, sometimes, unions cared less about students than they should have.

In addition, red-baiting during the McCarthy era, a broader structure of American labor relations that made it difficult for unions to engage in wider solidarities, and federal policies inscribing existing racial inequality served as powerful limits on the progressive possibilities for American teachers. Two examples of teacher unionism, one hundred years apart, however, show that teacher organization is at its best when it is part of a larger social movement and when it can show how intimately related are teacher working conditions, student learning conditions, and social equality. Sixth-grade teacher and pioneering union organizer Margaret Haley and her colleagues, for example, connected a conscious feminism in the late nineteenth century with evidence of how the wealthiest's failure to pay their share of taxes in Chicago harmed the entire city.[15] Not only did an overwhelmingly female teaching force get higher pay and stable pensions as a result of the effort, but the education system in Chicago was the better for its increased revenue as well. A century later, Chicago teachers were again on the vanguard of social-movement unionism. In 2010, an insurgent slate of activists—the Congress of Rank-and-File Educators (CORE)—took over leadership of the CTU after contentious union elections. CORE activists forged a powerful connection with the Chicago community by opposing Mayor Richard Daley's Renaissance 2010 plan—inherited by Democrat Rahm Emmanuel in 2011—to close public schools concentrated in minority neighborhoods and reconstitute them as charters. In contract negotiations in 2012, the union demanded real improvements in the schools as well as salary increases. They were able to get the overwhelming support of parents, went on strike, and clearly emerged victorious. Since then, the CTU has only deepened its critique of neoliberalism in Chicago, and, like Haley and the CTF in the early twentieth century, has powerfully connected the city's power structure to the school system.[16]

The movement led by the CTU—forging the broadest possible coalition against those who benefit most from the privatization of public goods and a rigged market competition—increasingly seems like, not just one way, but the only way toward a more equal society. Only by bringing together teachers of all backgrounds, a conscious feminist assault on gender inequality, a deeper critique of spatial inequality in the nation's metropolitan areas, and a real commitment to ensuring that all children have access both to good schools and freedom from the chronic stress and hunger so often brought on by poverty will the classroom become a space that facilitates a fair chance at a secure and fulfilling life for everyone.

NOTES

INTRODUCTION

1. Murphy, "Militancy in Many Forms," 231; "Fewer Teacher Strikes Reported but Many Disputes Are Unsettled," *New York Times (NYT)*, September 7, 1976.

2. On the "long 1970s," see Schulman, *Seventies,* xi–xvii. On the conservative turn in the Republican Party and U.S. politics, see Fraser and Gerstle, *Rise and Fall of the New Deal Order*; Critchlow, *Conservative Ascendancy*; Phillips-Fein, *Invisible Hands*; Moreton, *To Serve God and Wal-Mart*; Schulman, *Seventies*; Burns, *Goddess of the Market*; Self, *All in the Family*; Rick Perlstein, *Nixonland* and *Invisible Bridge*; and Cowie, *Stayin' Alive*. On the shift within the Democratic Party, see Stein, *Pivotal Decade,* and Battista, *Revival of Labor Liberalism.*

3. National Center for Education Statistics, *Digest of Education Statistics,* 2013, table 106.20, "Expenditures of Educational Institutions, by Level and Control of Institution: Selected Years, 1899–1900 through 2012–13," http://nces.ed.gov/programs/digest/d13/tables/dt13_106.20.asp, accessed July 15, 2016.

4. Metzgar's *Striking Steel* gives the single most persuasive portrait of this shift.

5. Kaye, *The Fight for the Four Freedoms.*

6. National Center for Education Statistics, *Digest of Education Statistics,* 1990, 44, http://nces.ed.gov/pubs91/91660.pdf, accessed July 15, 2016.

7. See, for example, McCartin, *Labor's Great War.*

8. Gross, *Broken Promise,* 1–2.

9. Lichtenstein, *State of the Union,* chap. 3. Lichtenstein argues that Taft-Hartley "prefigured and codified much of labor's postwar retreat"; see p. 114.

10. Vinel, *Employee.*

11. Useful discussion of the many ways in which companies were able to circumvent the Wagner Act by the 1970s, as well as the failed effort to reform it, can be found in Cowie, *Stayin' Alive,* 288–96, and Levitan and Cooper, *Business Lobbies,* 123–26.

12. See Davis, *Prisoners of the American Dream,* 65. Davis points out that the presidential election of 1936—which featured the largest popular margin in history up to that point—shifted the landscape of American politics because, for the

first time, class position became a better indicator of party affiliation than race or ethnicity.

13. Steve Fraser and Gary Gerstle, "Introduction," in Fraser and Gerstle, *Rise and Fall of the New Deal Order,* xi.

14. I build here on the framework of Battista, *Revival of Labor Liberalism,* 1–26. Battista refers to the term *labor-liberal coalition*—an alliance between unions and the northern liberal faction within the Democratic Party—as being preferable to *labor-Democratic coalition.* I concur that this is a decidedly superior term, and I extend its use to refer to the roots of the political assumptions during this era. This term is also superior because the coalition—particularly by the 1960s—did not totally break down along party lines. As we will see in this study, there were liberal Republicans—such as Governor Raymond Shafer in Pennsylvania or Mayor John Lindsay of New York City—whose policy assumptions mostly aligned with labor liberalism.

15. Governor's Commission to Revise the Public Employee Law report, June 25, 1968, series 1, RG-16, carton 1, Governor's Commission for the Revision of the State's Public Employee Law Records, Pennsylvania State Archives, Harrisburg, PA.

16. McCartin, "'A Wagner Act for Public Employees,'" 123–48, and "Turnabout Years," 210–26.

17. In the nineteenth century, as Marjorie Murphy has shown, teaching became a predominantly female occupation in large part because of budget realities—women could be paid less for the same work than men—and also because of assumptions about women's natural aptitude for child care and moral instruction. See Murphy, *Blackboard Unions,* 12–13. Clifford, in *Those Good Gertrudes,* highlights the historical reach of the assumptions regarding female teachers' special aptitude for child care as well as the assumptions undergirding a dramatically unequal system of remuneration for male and female teachers. In 1961, according to the National Center for Education Statistics, 68.7 percent of teachers were female. This number declined as teacher salaries increased in the late 1960s—to 65.7 percent in 1971—but then increased again in 1976 to 67 percent and, by 1986, was back to 68.8 percent; see *Digest of Education Statistics,* 1990, 77, table 63.

18. Seth King, "24,000 Teachers Strike, Closing 666 Schools," *NYT,* September 4, 1975; "2 Million Students across U.S. Idled by Teacher Strikes," *NYT,* September 11, 1975; "Teachers Jailed in '75," *Boston Globe,* September 25, 1975; Grace Lichtenstein, "Big City Problems in Big Sky Country," *NYT,* September 4, 1975; Lawrence Fellows, "Unions Pledge General Strike to Back New Haven Teachers," *NYT,* November 22, 1975.

19. Casey Banas and Stevenson Swanson, "City Teachers Poised to Strike," *Chicago Tribune,* September 8, 1987.

20. On the emergence of the "labor problem," see Stromquist, *Reinventing "The People."*

21. Sugrue, *Origins of the Urban Crisis*; Biondi, *To Stand and to Fight*; Self, *American Babylon*; Countryman, *Up South*; Freund, *Colored Property*; Gordon, *Mapping Decline*; and Highsmith, *Demolition Means Progress.*

22. Rubio, *There's Always Work at the Post Office.*

23. Hower, "Jerry Wurf, the Rise of AFSCME," chap. 5. On the link between AF-SCME and civil rights in the sanitation workers campaign, see also Honey, *Going Down Jericho Road.*

24. Piven, "Militant Civil Servants," 24–28; quotation is on p. 24.

25. Ture and Hamilton, *Black Power,* 164–71; quotation is on p. 166. It is important to point out that for Ture and Hamilton, in the context of the institutional racism of postwar schools, "professionalism" had been inherently racialized. Just having some black teachers and administrators was not enough to ensure the institutions would actually be accountable to African Americans. In their effort to characterize the United States as a colonial power that subjugated blacks, for instance, Ture and Hamilton argued there was "an entire class of 'captive leaders' in the black communities. These are black people with certain technical and administrative skills who could provide useful leadership roles in the black communities but do not because they have become beholden to the white power structures. These are black school teachers, county agents, junior executives in management positions with companies, etc"; see p. 13. Ture and Hamilton argued instead that entirely new black-led structures needed to be built, and only in that context could black teachers truly respond to the needs of black students.

26. Sugrue, *Origins of the Urban Crisis;* Self, *American Babylon;* the single best examination of deindustrialization over the course of the twentieth century is Cowie's study of the electronics industry, *Capital Moves.*

27. Stein, *Pivotal Decade,* especially chap. 7 and 8.

28. For an account of public views about AFDC—and the efforts of welfare rights activists to shape it, see Nadasen, *Welfare Warriors.*

29. Reese, *America's Public Schools,* 2.

30. Majority opinion, *Oliver Brown, et al., v. Board of Education of Topeka, et al.* (1954).

31. Rousmaniere, *Citizen Teacher.*

32. Murphy, *Blackboard Unions,* 82; Rousmaniere, *Citizen Teacher,* 159. The quotation, from 1915, was attributed to Chicago Board of Education member Jacob Loeb.

33. On the fears of the decline of the male breadwinner and what he calls the rise of "breadwinner conservatism," see Self, *All in the Family.* Self points out that "the single most dramatic development in the U.S. labor force between 1945 and 1970 was the increasing market presence of married women and women with children. By the early 1970s, more than half of all women between the ages of eighteen and sixty-four engaged in market work, and women constituted nearly 40 percent of all paid workers. . . . By 1974, 40 percent of all women with children between ages three and five were engaged in market work. This gradual resort to market labor by women was part of a decades-long trend, but the acceleration of that process in the 1960s was unmistakable"; see p. 110. Another instrumental study on this topic is Zaretsky, *No Direction Home.*

34. Kessler-Harris, *In Pursuit of Equity,* 7.

35. Sociologist David Halle's research on chemical workers in New Jersey in the early 1980s has shown that male blue-collar workers often resented the teachers who had sometimes made them feel unintelligent when they were going to school as teenagers. Although I found no direct evidence for this phenomenon, it seems likely that there was a connection between these kind of feelings and broader political characterizations of unionized teachers for some of the working-class men featured in this study. See Halle, *America's Working Man*, 48–50.

36. Johnston, *Success While Others Fail*, 4.

37. Cowie, *Stayin' Alive,* chap. 1; Cal Winslow, "Overview: The Rebellion from Below, 1965–1981," in Brenner, Brenner, and Winslow, *Rebel Rank and File*, 1–35.

38. Williams, *Public School and Finances*, 12–13; National Center for Education Statistics, *Digest of Education Statistics,* 1990, 147, table 147.

39. National Center for Education Statistics, *Digest of Education Statistics*, 2014, table 235.10, "Revenues for Public Elementary and Secondary Schools, by Source of Funds: Selected Years, 1919–20 through 2011–12," http://nces.ed.gov/programs/digest/d14/tables/dt14_235.10.asp?current=yes/, accessed July 15, 2016.

40. Wong, *Funding Public Schools*, 7.

41. Stein, *Pivotal Decade*, 176–261; Cowie, *Stayin' Alive,* 261–66.

42. Duménil and Lévy, *Capital Resurgent*, 1–2.

43. Phillips-Fein, *Invisible Hands*, chap. 2 and 3.

44. Milton Friedman, *Capitalism and Freedom*, 2, 6.

45. On his argument regarding the Great Depression, see Friedman and Schwartz, *Monetary History*, chap. 7 and 8. On Friedman's support for vouchers, see Carl, *Freedom of Choice*, 15–17. Carl argues that Friedman "was the public figure most responsible for infusing school voucher policies with intellectual credibility" (16).

46. Harvey, *Brief History of Neoliberalism*, 19, and *New Imperialism*, 149–50.

47. Fraser, *Age of Acquiescence*, 250.

48. Lipman, *New Political Economy of Urban Education,* 29–38; quotation is on p. 29.

49. Shermer, *Sunbelt Capitalism*, 3.

50. Robertson, "'Remaking the World,'" 13–16.

51. Harvey, *Brief History of Neoliberalism,* 2.

52. Perhaps the most illustrative visions for this kind of realignment can be found in a book titled *The Making of the New Majority Party,* written in 1975 by William Rusher—publisher of William F. Buckley's conservative magazine the *National Review*. The book-length argument for a third party centered on a history lesson: the New Deal, Rusher asserted, had shifted the axis of American politics toward a division between the "haves" and the "have-nots." In response to the Great Depression, the party of Franklin D. Roosevelt had realigned U.S. politics by setting apart "debtors from creditors, employees from employers, [and] the poor from the (less comparatively) well to do. A description of the major economic actions of the Roosevelt administration is little more than a list of steps taken by government to favor the former over the latter." Rusher argued that by the 1970s, however, political allegiances had shifted. Refashioning a theme that harked back at least as far as the

struggles of farmers and laborers to limit the purchase of new corporate power over their economic lives during the Gilded Age, Rusher asserted that the new "division pits the producers—businessmen, manufacturers, hard-hats, blue-collar workers and farmers—against a new and powerful class of non-producers comprised of a liberal verbalist elite (the dominant media, the major foundations and research institutions, the educational establishment, the federal and state bureaucracies) and a semi-permanent welfare constituency, all coexisting happily in a state of mutually sustaining symbiosis." Believing the Republican Party hopelessly beholden to its liberal wing, Rusher argued that a third-party movement was necessary to reap the gains of this "new majority" of "producers," and he wanted former California governor Ronald Reagan to lead the effort; see Rusher, *Making of the New Majority Party*, xx–xxi, 18, 115–17.

Historical work has documented the rise of "producerist" politics in the nineteenth century, although this version was certainly different from the iteration I document in this study. See, for instance, Fraser, *Age of Acquiescence*, which shows how both farmers and workers trained their criticism on emergent corporate power, and White, *Railroaded*, which shows how the producerist idiom in the Gilded Age was couched in the language of "antimonopolism." For an examination of the connection between producerism and populist politics during the course of the nineteenth and twentieth centuries, see Kazin, *Populist Persuasion*.

53. Harvey, *Brief History of Neoliberalism*, 50.

54. Lipman, *New Political Economy of Urban Education*, 12–13.

55. Mirel, *Rise and Fall of an Urban School System*.

56. Podair, *Strike That Changed New York*. See also Golin, *Newark Teacher Strikes*; Daniel Perlstein, *Justice, Justice*; Perrillo, *Uncivil Rights*; and Lyons, *Teachers and Reform*.

57. Murphy's *Blackboard Unions* is national in scale but predominantly deals with the geographic centers of teacher unionization: Chicago and New York.

CHAPTER I. "A NEW ERA OF LABOR RELATIONS"

1. Walter Goodman, "Why Teachers Are Striking," *Redbook*, March 1969, 67. Italics in original.

2. Public-opinion polls show widespread approval for labor unions during this era. When asked in a Gallup poll in 1953, "In general do you approve or disapprove of labor unions?," 75 percent of those asked approved while only 18 percent disapproved. In 1965, 71 percent of Americans approved while only 19 percent disapproved; *Gallup Poll: Public Opinion, 1935–1971*, 1590, 1592. Phillips-Fein's *Invisible Hands* has documented postwar corporate antiunionism in the 1950s and '60s, and Witwer's *Shadow of the Racketeer* has shown how journalist Westbrook Pegler dramatized labor corruption in the same period.

3. Public-opinion polls bear out this ideal: in a Gallup poll from December 1959— just after the settlement of a 116-day steel strike—59 percent of Americans favored a proposal to "settle labor disputes before nationwide strikes can start: If the unions

and the companies can't reach an agreement, then a special court would hand down a decision which both sides would have to accept." Only 21 percent opposed the proposal. In a January 1968 poll, 69 percent of Americans supported binding arbitration for strikes over twenty-one days; *Gallup Poll: Public Opinion 1935–1971*, 1643, 2112.

4. Ibid., 2178.

5. Cowie, *Stayin' Alive*; Lichtenstein, *State of the Union*, chap. 6; Moody, *Injury to All*.

6. Joseph McCartin has made this case in convincing fashion; see McCartin, "Bringing the State's Workers In." Other notable accounts of public-sector workers—in addition to the studies of teacher unions cited in the introduction—include Fink and Greenberg, *Upheaval in the Quiet Zone*; Freeman, *In Transit* and *Working-Class New York*, chap. 12–15; Ryan, *AFSCME's Philadelphia Story*; Boris and Klein, *Caring for America*; and Hower, "Jerry Wurf, the Rise of AFSCME."

7. Leo Troy, "The Rise and Fall of American Trade Unions: The Labor Movement from FDR to RR," in Lipset, *Unions in Transition*, 80–84.

8. Clifford, *Those Good Gertrudes*, 1–44; William Reese, *America's Public Schools*, 28–44. For an example of the important role many believed women held in instructing children in the republican values of the early national period, see Benjamin Rush's essay "Thoughts upon Female Education," published in 1787, at http://quod.lib.umich.edu/e/evans/N16142.0001.001/1:1?rgn=div1;view=fulltext, accessed July 15, 2016.

9. Murphy, *Blackboard Unions*, 61–79; Rousmaniere, *Citizen Teacher*, 44–111.

10. Murphy, *Blackboard Unions*, 84–86; Rousmaniere, *Citizen Teacher*, 175.

11. Murphy, *Blackboard Unions*, 109–21.

12. See Roosevelt's letter on the Resolution Federation of Federal Employees against Strikes in Federal Service on August 16, 1937, at http://www.presidency.ucsb.edu/ws/?pid=15445, accessed July 15, 2016.

13. Murphy, *Blackboard Unions*, 91–92.

14. Taylor, *Reds at the Blackboard*, chap. 3.

15. Murphy, *Blackboard Unions*, 131–32, 138–40, 150–74; Taylor, *Reds at the Blackboard*, chap. 1–3.

16. Murphy, *Blackboard Unions*, 175–95.

17. Gaffney, *Teachers United*, 32–33.

18. Ibid., 33; Murphy, *Blackboard Unions*, 215.

19. Taylor, *Reds at the Blackboard*, chap. 6 and 7.

20. Murphy, *Blackboard Unions*, 185–95.

21. Taylor, *Reds at the Blackboard*, 3. Clearly, red-baiting limited the political possibilities for teachers in the postwar era. It is also possible, as Zoë Burkholder has argued, that the focus in the historiography on left-wing unions has caused historians to miss the fact that a significant bloc of New York teachers—even in the early 1940s—were already predisposed to political conservatism; see Burkholder, "'A War of Ideas.'"

22. Perrillo, *Uncivil Rights*, 4–7 and chap. 2 and 3.

23. Gaffney, *Teachers United,* 10.

24. Kahlenberg, *Tough Liberal,* 33.

25. Lortie, *Schoolteacher,* 11, 82–108; quotation is on p. 11.

26. Gaffney, *Teachers United,* 30–31.

27. On the gendered notions of "economic citizenship" in the twentieth century challenged by resurgent feminism, see Kessler-Harris, *In Pursuit of Equity.* On the attempts on the right to defend traditional gender roles in the guise of "breadwinner conservatism" during the period of this study, see Self, *All in the Family.*

28. Murphy, *Blackboard Unions,* 196–200, 206–7.

29. Lyons, *Teachers and Reform,* 155–56.

30. Murphy, *Blackboard Unions,* 215–18.

31. Selden, *Teacher Rebellion,* 47–86; Murphy, *Blackboard Unions,* 192–211; quotation is on p. 209.

32. Friedman, "Role of the AFL-CIO," 30–33. On organized labor and Detroit teachers, see Mirel, *Rise and Fall of an Urban School System,* 313–16.

33. Murphy, *Blackboard Unions,* 220; "Louisville Teachers Picket and File Suit in Strike," *New York Times (NYT),* November 11, 1964; "Teachers' Strike in South Bend Continues in Face of Dismissals," *NYT,* May 15, 1965; Leonard Buder, "Teachers' Strike Averted but Questions Are Raised," *NYT,* September 12, 1965; Leonard Buder, "A Warning to Strikers," *NYT,* April 3, 1966; Walter Waggoner, "Newark's Schools Remain Open Despite Strike by 750 Teachers," *NYT,* December 3, 1965; "Ten Newark Teachers and Union Are Fined $3700 for Contempt," *NYT,* December 24, 1965.

34. "Newark Teachers End 2-Day Strike; Accept Arbitration," *NYT,* February 14, 1966; Leonard Buder, "A Warning to Strikers," *NYT,* April 3, 1966; "Teachers Boycott Michigan Schools," *NYT,* June 3, 1966.

35. Murphy, *Blackboard Unions,* 220; "Teachers in Baltimore Arrested for Picketing," *NYT,* May 12, 1967; "Strikes Halt Fall Classes across U.S.," *Chicago Tribune (CT),* September 6, 1967; "Teacher Strikes, Disputes Hit 6 States," *CT,* September 7, 1967; Donald Janson, "Teachers' Talks Stall in Michigan," *NYT,* September 8, 1967.

36. Mirel, *Rise and Fall of an Urban School System,* 111–24, 171–86, 271–73, 313–20; "230 Teachers in Michigan Told 'Back-to-Work'" *CT,* September 10, 1967.

37. Freeman, *Working-Class New York,* 208–11.

38. Gaffney, *Teachers United,* 34–38.

39. Ibid., 38–39.

40. Peter Millones, "Taylor Law's Baptism," *NYT,* September 27, 1967.

41. Tom Wicker, "In the Nation: Trouble in the Schoolhouse," *NYT,* September 10, 1967.

42. "Chicago Teachers Accept Agreement," *NYT,* January 8, 1968.

43. "Cincinnati Strike Ends," *NYT,* February 2, 1968; "Leaders of Teachers Union Jailed for Cincinnati Strike," *NYT,* February 16, 1968.

44. Fred Hechinger, "Florida Strike Is a Test for the Militant Teacher," *NYT,* February 25, 1968. Estimates of the total number of teachers on strike varied. The first accounts

indicated that 35,000 teachers resigned, but later estimates place the numbers as low as 22,000. See Associated Press, "Teachers Out in Florida," February 16, 1968, and "Teachers' 'Revolution' Ripples through Nation," February 20, 1968. On the school closings and the end to the strike, see Associated Press, "Legislators Shun Meeting," February 26, 1968, and "Fla. School Strike Pact Hits Snag," March 11, 1968.

45. Letter from Carl Megel to John Livingston, and Megel to Henry McFarland, RG28–002, box 47, AFL-CIO Collection, George Meany Archives, University of Maryland, College Park.

46. "Teacher Cites Hard Road, Low Pay," *Pittsburgh Press* (*PP*), February 20, 1968; "Teachers Figured Working 50 Hours," *PP*, February 27, 1968.

47. "Teachers' Demands Judged Unfairly," *Pittsburgh Post-Gazette* (*PPG*), February 22, 1968; "Teachers' Protests Endorsed," *PP*, February 18, 1968.

48. "Striking Teachers Close 13 Schools," *PPG*, March 1, 1968.

49. L. R. Lindgren, "Shafer Doing Juggling Act," *PP*, February 25, 1968; Jack Ryan, "City, State Will Press to End School Strike," *PPG*, March 9, 1968; "City Teachers End 11-Day Strike," *PPG*, March 11, 1968.

50. "The Teachers' Threat," *PP*, February 20, 1968.

51. Cy Hungerford, "Contagious?," *PPG*, February 22, 1968.

52. Cy Hungerford, "Thursday Walkout," *PPG*, February 26, 1968.

53. "Teachers' Protest Seminar Called Harmful to Children," *PP*, March 2, 1968.

54. Letter to the editor, *PP*, March 5, 1968.

55. Letter to the editor, *PPG*, March 2, 1968.

56. Mills, *Power Elite*, *White Collar*, and *New Men of Power*; Whyte, *Organization Man*.

57. "The Teachers' Threat," *PP*, February 20, 1968.

58. Editorial, *PP*, March 4, 1968.

59. "State Income Tax, Teacher Pay Hike Linked by Shafer," *PPG*, March 29, 1968.

60. Cy Hungerford, "The New Math," *PPG*, March 29, 1968.

61. "City Schools Seek Additional Funds," *The News from the Pittsburgh Public Schools*, October–November 1970, p. 1, Pittsburgh Board of Public Education, series IV, box 182, folder 1, Heinz Historical Center Archives, Pittsburgh.

62. "Teachers Move to Higher Pay Schedule," *The News from the Pittsburgh Public Schools*, October–November 1972, p. 9, Pittsburgh Board of Public Education, series IV, box 182, folder 1, Heinz Historical Center Archives, Pittsburgh.

63. Jack Ryan, "Irvis, Lamb Will Continue to Offer Aid," *PPG*, March 9, 1968.

64. Memo from Raymond Shafer to the State Education Secretary, May 14, 1968, RG-16, carton 1, Pennsylvania State Archives, Harrisburg, PA (hereafter, PSA).

65. On Detroit, see Mirel, *Rise and Fall of an Urban School System*, 378, n. 3.

66. Hickman Commission Report and Recommendations, presented to the governor's office, June 25, 1968, RG-16, carton 1, PSA.

67. Ibid.

68. Schlesinger viewed the cornerstone of American democracy as the ability of different interests to prevent each other from controlling the state. Thus, the public

interest, according to Schlesinger in 1945, could only be served by negating the influence of powerful groups: "The business community has ordinarily been in the most powerful of these groups, and liberalism in America has been ordinarily the movement on the part of the other sections of society to restrain the power of the business community." Galbraith pointed out, several years later, that labor unions should serve as a "countervailing power" to "push back" against the interests of corporations. Schlesinger, from *The Age of Jackson* (1945), and Galbraith, from *American Capitalism* (1952), are both quoted in Mattson, *When America Was Great*, 98, 101–2. The participation of both Galbraith and Schlesinger as political appointees in the Kennedy administration shows how close this position was to the political mainstream by the 1960s. President Kennedy would show his commitment to the notion that unions provided counterweight to business interests by elevating United Steelworkers of America general counsel Arthur Goldberg first to secretary of labor and then, in 1962, to the Supreme Court.

69. The law not only received every Democratic vote in the state House of Representatives, but also passed a state senate holding a Republican majority.

70. Hawaii passed a law allowing public-sector strikes just a month before, and Vermont already allowed public-sector strikes in more restrictive circumstances.

71. "Fondy Points Out Progress in New Public Employee Law," *PP*, July 24, 1970.

72. William E. Deibler, "Shafer Okays Law on Public Employees," *PPG*, July 23, 1970.

73. Ibid.; Vincent Carocci, "Shafer Approves Public Employees' Strike-Right Law," *Philadelphia Inquirer (PI)*, July 23, 1970. Shafer quotations are from the Carocci article.

74. "New Public Employee Law," *PPG,* July 22, 1970.

75. "Right to Strike: It Carries Responsibilities, Too," *Harrisburg Patriot,* July 25, 1970.

76. "Teacher Strikes across Nation Delaying School for Thousands," *NYT,* September 5, 1968.

77. Donald Janson, "Teachers' Strike Halted in Chicago," *CT,* May 23, 1969; Anthony Ripley, "Denver Teachers Vote to Accept Pact Ending Two-Week Walkout," *NYT,* December 1, 1969; John Fenton, "Teachers Accept Providence Pact," *NYT,* December 12, 1969; Walter Waggoner, "Teachers Ratify Jersey City Pact," *NYT,* March 5, 1970; "Teachers Leader Jailed, Union Fined in Boston," *NYT,* May 7, 1970.

78. Harry Bernstein, "Teachers' Strike Cripples Schools," *Los Angeles Times (LAT)*, April 14, 1970; Harry Bernstein and Jack McCurdy, "Nearly 65% of Teachers Strike, Leaders Claim," *LAT,* April 15, 1970; Donovan, "Tale of Two Strikes."

79. "Teachers' Strike Views," *LAT,* April 11, 1970.

80. Rick Perlstein, *Invisible Bridge*, 408–9.

81. For the definitive account of PATCO, see McCartin, *Collision Course.*

82. Ronald Reagan, "Remarks and a Question-and-Answer Session with Reporters on the Air Traffic Controllers Strike, August 3, 1981," http://www.reagan.utexas.edu/archives/speeches/1981/80381a.htm, accessed July 15, 2016.

83. Harry Bernstein, "Teachers' Union Plans to Intensify Walkout," *LAT,* April 20, 1970.

84. Fred Hechinger, "A Money Squeeze Marks the Opening of Schools," *NYT,* September 13, 1970.

85. "The Planned Teachers' Strike," *LAT,* April 9, 1970.

86. John Corr, "School System's Plight: Desperate," *PI*, February 6, 1970.

87. Countryman, *Up South,* 50–56.

88. The starting salary for Philadelphia teachers in 1969–70 ranged from $7,300 to $7,600; Chicago's lowest starting salary was $8,400; New York City, $8,450; Detroit, $7,616; Washington, DC, $8,000; and Baltimore and Cleveland, $7,000. The number of years required to reach the top salary step in Philadelphia—ten years—was the same as that in both Chicago and Detroit. See "Teachers' Pay Compared," *Philadelphia Evening Bulletin (PEB)*, September 9, 1970.

89. "Still Passing By on the Other Side," *PI*, February 9, 1970.

90. John T. Gillespie, "Board Orders School Closed as Pact Ends," *PEB*, September 8, 1970.

91. John T. Gillespie, "City Pupils Return Tomorrow under 30-Day Agreement," *PEB*, September 14, 1970.

92. John Corr, "Board, Union Reach Settlement; All Teachers Return to Classes," *PI*, October 21, 1970.

93. Letter from James H. J. Tate to Frank Sullivan, PFT Contract Files, box 163, Philadelphia Federation of Teachers Collection, Urban Archives, Temple University Library, Philadelphia, PA (hereafter, PFTC/UA).

94. "The Governor and Philadelphia's School Crisis," August 28, 1970, WCAU-TV editorial, PFT Contract Files, box 163, PFTC/UA.

95. Letter from Gail R. Aronson to Frank Sullivan, October 7, 1970, PFT Contract Files, box 163, PFTC/UA.

96. "A Strike and a Wrong One," *PEB*, September 9, 1970.

97. "Billion for '76, but for Schools . . . ," *PI*, October 19, 1970.

98. Letter from Alexander Layman to Frank Sullivan, September 12, 1970, PFT Contract Files, box 163, PFTC/UA.

99. Letter from Mr. and Mrs. Irv Forman, October 12, 1970, PFT Contract Files, box 163, PFTC/UA.

100. Letter from Dorothy Ravich to Frank Sullivan, September 29, 1970, PFT Contract Files, box 163, PFTC/UA.

101. "Illegal and Irresponsible," *PI*, October 19, 1970.

102. "14 Union Leaders Are Jailed in New Haven Teacher Strike," *NYT,* September 18, 1970; "Hartford Teachers End Their Strike," *NYT,* December 6, 1970.

103. Vince Gagetta, "City Schools Resume Normal Schedule after PFT Strike," *PPG*, January 18, 1971.

104. Ibid.

105. The next contract, incidentally, was negotiated without a strike.

106. "School-Strike Aftermath," *PP*, January 18, 1971.

107. "We Are Glad to Be Back!," *PPG*, January 18, 1971.

108. See *NBC Nightly News,* January 5 and January 6, 1971, Television News Archive, Vanderbilt University, Nashville, TN.

CHAPTER 2. TEACHER POWER, BLACK POWER, AND THE FRACTURING OF LABOR LIBERALISM

1. Charles Hamilton, "An Advocate of Black Power Defines It," *NYT,* April 14, 1968.
2. Ture and Hamilton, *Black Power,* 167–71.
3. Charles Hamilton, "An Advocate of Black Power Defines It," *NYT,* April 14, 1968.
4. Perrillo, *Uncivil Rights,* chap. 3.
5. Podair, *Strike That Changed New York,* 50–58.
6. Ibid., chap. 4.
7. Ibid., 17–20.
8. Ibid., chap. 5–6.
9. Ibid., chap. 3; quotation is on p. 69.
10. Daniel Perlstein, *Justice, Justice,* chap. 1, 2, 7, and 8; quotations are on pp. 23, 8. For a particularly useful account of Shanker's view of "race-blind" merit, or what he calls "color-blind" tough liberalism, see Richard Kahlenberg's biography of the UFT/AFT president, *Tough Liberal,* 1–11.
11. Perrillo, *Uncivil Rights,* 118–19. Jane Anna Gordon's 2001 treatment of Ocean Hill–Brownsville represents the first major effort to view community control as a "sensible" response to the grossly unequal conditions of education in the schools; see Gordon, *Why They Couldn't Wait.*
12. Podair, *Strike That Changed New York,* 91; Earl Caldwell, "Negroes and Pickets Clash at School," *NYT,* September 13, 1967. Richard Kahlenberg points out that MES—the brainchild of Shanker—represented, to the UFT president, the nation's "first magnet schools"; see Kahlenberg, *Tough Liberal,* 57.
13. "Text of the Decision by Nunez in Shanker's Trial," *NYT,* October 5, 1967.
14. "Road to Civic Chaos," *NYT,* September 11, 1967.
15. "Shanker Conviction in Strikes Upheld," *NYT,* June 6, 1969; Damon Stetson, "Taylor Law Held Failure by Kheel," *NYT,* October 13, 1968.
16. Podair, *Strike That Changed New York,* 116, 174–82; Daniel Perlstein, *Justice, Justice,* chap. 4.
17. Damon Stetson, "A.F.L.-C.I.O. Backs Teachers," *NYT,* September 7, 1968; Podair, *Strike That Changed New York,* 112, 134.
18. Jones, *March on Washington,* chap. 4.
19. "An Appeal to the Community from Black Trade Unionists," *New York Amsterdam News,* September 28, 1968.
20. Peter Millones, "Union Solidarity Is Shaken by Strike," *NYT,* October 24, 1968; "Negro Unionists Back Ocean Hill," *NYT,* October 26, 1968; Podair, *Strike That Changed New York,* 134.
21. Resnik, *Turning on the System,* 42, 61.

22. Countryman, *Up South,* chap. 6. The PFT response to the demonstration is noted on p. 243.

23. "Integration Set in Philadelphia," *NYT,* December 22, 1967; "N.Y. School Strike over Black Rule," *New Pittsburgh Courier,* September 28, 1968; "Students Angry in Philadelphia," *NYT,* November 16, 1969; "Disturbing Questions Left by Recent Teachers Strike," *Philadelphia Tribune,* October 24, 1970.

24. Mirel, *Rise and Fall of an Urban School System,* 294–98; quotation is on p. 295.

25. Ibid., 298–313; quotation is on p. 309.

26. Ibid., 320, 336.

27. Lyons, *Teachers and Reform,* 44, 74–75, 96, 134.

28. Ibid., 135–37.

29. Ibid., 149–54, 175, 184–98.

30. Ibid., 198–204.

31. The only in-depth scholarly treatment of the Newark conflict is Steve Golin's *Newark Teacher Strikes.*

32. Ibid., 18–19.

33. Golin, *Newark Teacher Strikes,* 32–71; quotation is on p. 38.

34. Ibid., 72–75.

35. Robert Braun, "Kheel Entering Newark Strike; 3 Teacher Leaders Surrender Today," *Newark Star-Ledger (S-L),* February 5, 1970.

36. Golin, *Newark Teacher Strikes,* 40–42; Woodard, *Nation within a Nation,* 75. Although African Americans made up just 17 percent of the city's population in 1950, by 1970, the city was 54 percent black.

37. Mumford, *Newark,* chap. 5–6.

38. Woodard, *Nation within a Nation,* chap. 3–4; Mumford, *Newark,* chap. 9.

39. Robert Braun, "LeRoi Jones: His 'Aides' Help Keep Schools Open," *S-L,* February 8, 1970.

40. Robert Braun, "Teacher Strike: Frustration . . . Fear . . . Distrust," *S-L,* February 22, 1970.

41. Golin, *Newark Teacher Strikes,* 99–102; Robert Braun, "New Contract Ratified; Teachers Returning Today," *S-L,* February 26, 1970.

42. Golin, *Newark Teacher Strikes,* 124–25.

43. Imamu Amiri Baraka, "Crisis in Newark," *NYT,* April 16, 1971.

44. Robert Braun, "Teacher Talks: Indirect Openers," *S-L,* February 4, 1971.

45. Robert Herbert and Robert Braun, "NTU Strike Shows Sign of a Break," *S-L,* February 18, 1971.

46. Robert Braun, "NTU Chiefs Silent on Moves," *S-L,* February 26, 1971; Robert Braun, "Union Leaders Refuse to Order Teachers Back," *S-L,* March 1, 1971.

47. Imamu Amiri Baraka, "White Suburban Teachers vs. Black NewArk," CFUN newsletter. Newsletter has the handwritten date of "7–17–71," but this is possibly the date on which it was received by the AFT headquarters; American Federation of Teachers President's Office Records, box 13, Walter P. Reuther Library, Wayne State University, Detroit, MI (hereafter AFTPOR).

48. Robert Braun, "School Strike Settlement Hangs on a Few Points," *S-L*, March 22, 1971.

49. Roger Witherspoon, "Jacob Criticizes Turco for Supporting Teachers," *S-L*, March 2, 1971; "Turco Demands Resignations of Jacob, Saunders," *S-L,* March 3, 1971.

50. Golin, *Newark Teacher Strikes,* 140–81; quote on p. 144. For a fascinating treatment of the tête-à-tête between Imperiale and Baraka, see Mumford, *Newark,* chap. 8.

51. "Newark—Teacher Strike," *CBS Evening News,* February 11, 1971, Television News Archive, Vanderbilt University, Nashville, TN (hereafter, TNA).

52. "Newark, NJ: Teacher Strikes," *ABC Evening News,* April 8, 1971, TNA.

53. Robert Braun, "NTU Offer Gives Impetus to Talks," *S-L*, March 25, 1971.

54. Lawrence Hall, "Blacks, Whites Polarized at Board Meeting," *S-L*, April 7, 1971.

55. "Finley, Board Members Reflect on Their Crucial Balloting," *S-L*, April 8, 1971.

56. Charles Finley and James Amanna, "Parents, Teens Protest at City Hall," *S-L,* April 11, 1971.

57. "Strikes: Newark, NJ," *ABC Evening News,* April 16, 1971, TNA.

58. Charles Finley and Nancy Jaffer, "Board, Teachers OK Pact," and Nancy Jaffer, "End of Strike Greeted with Relief and Regret," *S-L*, April 19, 1971.

59. Zaretsky, *No Direction Home.*

60. Dorothy Gallagher, "The Teacher Who Chose to Go to Jail," *Redbook,* January 1971, 58–59, 160–64.

61. "Schoolteachers/Jail," *NBC Nightly News,* December 23, 1971; "Teachers' Jail Release," *NBC Nightly News*, December 31, 1971, TNA.

62. A fuller discussion of the Selden letters can be found in Shelton, "Letters to the Essex County Penitentiary."

63. Selden, *Teacher Rebellion,* 4.

64. Witwer, "Heyday of the Labor Beat."

65. Robert Braun, "Whoever Wins the Bitter Power Struggle over Newark Schools . . . the Children Lose," *S-L*, March 28, 1971. See also Robert Braun, "Disastrous Impact on Pupils: School Strike Traumatic," *S-L,* March 7, 1971.

66. Milton Friedman's *Free to Choose* series, aired on PBS in 1980 (see chapter 6), is an example of this farther-reaching attack.

67. Braun, *Teachers and Power,* 11, 16.

68. Ibid., 260, 276.

69. Squires, *Read All about It!*

70. See Office of the President Collection, box 12, AFTPOR.

71. Letter from R. E. Moore, April 23, 1970, box 12, AFTPOR. A California high school student offered that "reading about your arrest gave me yet another reason not to be proud of America. . . . Your requests are reasonable and justified. Court injunctions and arrests are not"; letter from Paula Mather, undated, box 12, AFTPOR.

72. Check sent from James Arbuckle, April 18, 1970; check sent from Dave Becker, April 19, 1970; unsigned, undated defacing of AFT advertisements; un-

signed, undated letter; letter from Al Eischen, April 20, 1970; letter from Fred Knolldoff, April 20, 1970; letter from Edward Withing, April 15, 1970; all box 12, AFTPOR.

73. Perlstein, *Nixonland*, 746–47; italics in original.

74. Letter from David Perry, April 15, 1970, box 12, AFTPOR.

75. Letter from Ralph Curcio, April 15, 1970, box 12, AFTPOR; underlining in original.

76. Letter from John Amber, April 17, 1970, box 12, AFTPOR.

77. O'Connor, "Financing the Counterrevolution."

78. Letter from Barbara Mancbach, April 14, 1970, box 12, AFTPOR.

79. Reese, *America's Public Schools*, 3.

80. Countryman, *Up South,* 179.

81. Carter, *Politics of Rage*, 347–51.

82. Ibid.; Rick Perlstein, *Nixonland*, 223–24, 433–35.

83. Defaced advertisement from *New York Times*, April 19, 1970, box 12, AFTPOR; underlining in original.

84. Unsigned, defaced advertisement from the *New York Times,* April 18, 1970, box 12, AFTPOR.

85. Unsigned, defaced advertisement from the *New York Times,* April 19, 1970; letter from Edward Withing, April 15, 1970; letter from Jack Sherman, April 20, 1970; all box 12, AFTPOR.

86. Defaced *Chicago Tribune* advertisement, April 16, 1970, box 12, AFTPOR.

87. Defaced *New York Times* advertisement, April 18, 1970, box 12, AFTPOR.

CHAPTER 3. "WHO IS GOING TO RUN THE SCHOOLS?"

1. U.S. Census Bureau, http://www.census.gov/population/www/documentation/ twps0027/tab20.txt, accessed July 15, 2016. St. Louis, had, in the 1960 census, been one of the ten largest cities in the United States.

2. *ABC Evening News,* February 2, 1973, Television News Archive, Vanderbilt University, Nashville, TN.

3. U.S. Department of Labor, Bureau of Labor Statistics, Databases, Tables, and Calculators by Subject, "Consumer Price Index—All Urban Consumers, 1958–2015," http://www.bls.gov/data/, accessed July 15, 2016.

4. National Center for Education Statistics, *Digest of Education Statistics,* 1990, 83, table 71, http://nces.ed.gov/pubs91/91660.pdf, accessed July 15, 2016.

5. See Fred Hechinger, "A Good Break for Most Teachers," *NYT,* August 29, 1971.

6. Charles F. Thomson, "Rizzo Draws Heat as Teachers Stay Out," *Philadelphia Evening Bulletin (PEB),* September 24, 1972; Linda Loyd and Thomas Hine, "City's Schools to Reopen Friday under Old Pact; Talks Continue," *Philadelphia Inquirer (PI),* September 28, 1972.

7. "Women Support General Strike," *Philadelphia Tribune (PT),* February 27, 1973.

8. See Aaron Epstein and Jon Katz, "Rizzo Weakens on Pledge; Has 'Plans for New Taxes,'" *PI,* February 28, 1973; Duke Kaminski, "Battle Seen in Senate over

Phila. School Aid," *PEB,* February 27, 1973. For example, Rizzo advocated taxes on pinball and vending machines.

9. September 11, 1972, board meeting, *Journal of the Board of Education of Philadelphia*, Philadelphia Free Library. Ross ended up resigning his seat on the school board, and the Philadelphia AFL-CIO censured him for opposing the teachers. The PFT unsuccessfully attempted to get Ross expelled from the Philadelphia AFL-CIO Executive Council. See letter from Frank Sullivan to George Meany, May 22, 1973, box 13, folder 13, AFL-CIO Office of the President files, George Meany Archives, University of Maryland, College Park.

10. Ryan, *AFSMCE's Philadelphia Story,* 1–9, 39–40.

11. WCAU-TV editorial, August 31, 1972, PFT Contract Files, box 173, Philadelphia Federation of Teachers Collection, Urban Archives, Temple University Library, Philadelphia, PA (hereafter, PFTC/UA).

12. Sterling and Haight, *Mass Media,* 335–37.

13. "Beyond the Strike," *PEB*, September 17, 1972. On the Chamber of Commerce, see Moreton, "Make Payroll, Not War."

14. See Creed Black, "State's Public Employee Law Isn't Working," *PI*, September 10, 1972.

15. "The Teachers Should Reconsider," *PI*, September 21, 1972.

16. George Wilson, "Still the Key Issue: Who Runs Schools?," *PI*, September 30, 1972.

17. Letter to Frank Sullivan, September 20, 1972, PFT Contract Files, box 173, PFTC/UA.

18. "Should Teachers Sacrifice Previous Gains?," "Wants to Teach Not Supervise," "Editorial Showed 'Tortuous' Logic," *PEB*, September 20, 1972.

19. "Strike Is over Working Conditions," *PEB*, September 22, 1972.

20. "Sick and Tired," *PEB*, September 20, 1972.

21. Letter from Jack J. Morris, September 21, 1972, PFT Contract Files, box 173, PFTC/UA.

22. "School First, Then More Police," "Divert Vietnam Billions to Schools," *PEB*, September 22, 1972.

23. "Pupils, Adults Blame Teachers for Strike in Phila. Schools," *Sunday Bulletin*, September 24, 1972.

24. Letter to Frank Sullivan, December 8, 1972, PFT Contract Files, box 173, PFTC/UA.

25. Letter from Joan Goldberg to Sullivan, Dec. 8, 1972. PFT Contract Files, box 173, PFTC/UA.

26. Resnik, *Turning on the System,* 61. See also Orfield, *Must We Bus?,* 62.

27. Thomas Hine and Linda Loyd, "Rizzo: We're Not a Bunch of Patsies," *PI*, September 27, 1972.

28. Letter to Frank Sullivan, September 17, 1972, PFT Contract Files, box 173, PFTC/UA; underlining in original.

29. Letter to Frank Sullivan, September 18, 1972, PFT Contract Files, box 173, PFTC/UA; Letters to the editor, *PI*, September 22, 1972.

30. "Teachers Act Like Longshoremen," *PEB*, September 26, 1972; "Striking Teacher Unworthy of Title," *PI*, September 26, 1972.

31. Survey at Stephen Decatur Home and School Association, November 1972, PFT Contract Files, box 173, PFTC/UA.

32. Letters to Frank Sullivan, November 28, December 6, and December 11, 1972, PFT Contract Files, box 173, PFTC/UA; all underlining in original.

33. Letters to Frank Sullivan, September 12, 1972, September 15, 1972, and September 16, 1972, PFT Contract Files, box 173, PFTC/UA. The September 16 letter featured an illegible last name, but her first name was Dorothy.

34. "Youths Protest against School Strike," *PT*, September 12, 1972; standalone photo, *PT*, September 23, 1972.

35. Anthony Astrachan, "Sketches of a Strike," *Washington Post*, March 4, 1973. As in Newark, a large portion of black teachers opposed the strike because they believed it undermined the interests of the city's predominantly African American school students. See, for example, the story of Carrie McFadden, the oldest of six in an African American family from North Philadelphia. "Not only am I against the union in the strike but I am against the board and the city administration too. If the racial makeup of the school system was reversed—predominately white instead of black—this wouldn't be happening"; see Acel Moore, "My Only Loyalty Is to the Children," *PI*, January 12, 1973.

36. Letters to Frank Sullivan, November 16, 1972, and December 12, 1972, PFT Contract Files, box 173, PFTC/UA.

37. January 5, 1973, board meeting, *Journal of the Board of Education of Philadelphia* (1972–73).

38. Ibid.

39. Jon Katz, "Report to School Monday, Rizzo Tells Teachers, Pupils," *PI*, January 6, 1973.

40. "Philadelphia Schools Do Concern Main Line," *PI*, January 3, 1973; "A Strike Won't Help," *PEB*, January 7, 1973.

41. "Teachers Have Their Say on Strike Threat," *PEB*, January 5, 1973; "Blame City Teachers for the School Crisis," *PI*, January 7, 1973.

42. Charles Thomson, "Judge Jamiesen Issues Injunction to End 4-Day Teachers' Strike," *PEB*, January 11, 1973; "For Philadelphia's Children 'A Clear and Present Danger,'" *PEB*, January 12, 1973.

43. "The Teachers Should Comply with the Law and the Court," *PI*, January 12, 1973; "For Philadelphia's Children 'A Clear and Present Danger,'" *PEB*, January 12, 1973.

44. "Hire the Dedicated," *PI*, January 14, 1973; "The Rule of Law," *PI*, January 16, 1973.

45. "Critics of Teachers' Strike Sound Off," *PEB*, January 26, 1973.

46. "Defying the Court," *PEB*, January 16, 1973; WCAU-TV editorial, January 18, 1973, PFT Contract Files, box 173, PFTC/UA.

47. "Sexist School Crisis," *PI*, January 17, 1973.

48. Charles Thomson, "Sullivan, Ryan Get 6-Month Jail Terms; 48 Labor Chiefs Threaten General Strike," *PEB*, February 9, 1973; "360 Seized at 9 Schools, Face Charges This Time," *PEB,* February 19, 1974.

49. Letter to John Ryan, February 6, 1973, and letter to Sunny Richman, postmarked February 6, 1973, PFT Contract Files, box 173, PFTC/UA. On February 10, the same writer, in unmistakably the same handwriting, wrote Richman again: "When you act like Jesse James you get treated like Jesse James. Your demands are outrageous, you are acting like stevedores."

50. "Suggestion to Avoid a Mass Strike," *PEB*, February 23, 1973; "Did Union Leaders Deserve Jail Terms?," *PI*, February 14, 1973.

51. "While Teachers Vote—a Sigh of Relief," *PEB*, February 28, 1973.

52. "Readers Study the Strike Lesson," *PEB*, February 4, 1973; "School Strike Showed Act 195 Worthless," *Harrisburg Patriot*, March 2, 1972, reprinted in *PI*, March 3, 1973.

53. Pamala Haynes, "Right On," *PT,* January 20, 1973.

54. "Philly School Problems Seen as Both Racial, Financial," *Atlanta Daily World,* March 1, 1973.

55. "Blames Union," *PT,* January 16, 1973.

56. Edith Herman and Ronald Yates, "Union Leaders Urge Strike, but Lower Demands on Wages," *Chicago Tribune* (*CT*), January 3, 1973; Edith Herman, "What Do Teachers Want? Pay Hike Only 1 Item on List," *CT,* January 7, 1973; "Who's Being 'Driven Out'?," *CT,* January 7, 1973; Don Papson, "The Union's Side," *CT,* January 10, 1973.

57. "St. Louis Teachers Vote to Strike over Starting Pay," *CT,* January 22, 1973.

58. "Victory for the Union," *CT,* January 26, 1973.

59. "Teacher Poll May Have Cooled Union," *Chicago Daily Defender* (*CDD*), January 6, 1973.

60. Letter to the editor, *CDD,* January 27, 1973.

61. "Teachers Strike Ends," *CDD,* January 27, 1973.

62. "PUSH Backing Teachers," *CDD,* January 8, 1973.

63. "The School Crisis," *CDD,* January 11, 1973.

64. "Bargaining Rights for Teachers," *CT,* September 25, 1965.

65. "Who's Being 'Driven Out'?," *CT,* January 7, 1973.

66. "The Board Finds Its Voice," *CT,* January 11, 1973.

67. "After a Short School Week," *CT,* January 13, 1973.

68. "Philadelphia Story," *CT,* March 7, 1973.

69. "Board Backed by Opinions," *St. Louis Post-Dispatch* (*SPD*), January 22, 1973.

70. DuBose press release, September 4, 1968, in American Federation of Teachers President's Office Records, Office of the President Correspondences, box 16, Walter P. Reuther Library, Wayne State University, Detroit, MI; letter from Jerry Abernathy and DuBose to Adella Smiley, December 22, 1972, in St. Louis Teachers Union Local

420 Records, box 18, Walter P. Reuther Library, Wayne State University, Detroit, MI (hereafter, STLTUC); Karen Van Meter, "Teachers Angered by High Pay for Craftsmen," *SPD,* February 11, 1973.

71. Meeting minutes, January 9, 1973, *Printed Record of the St. Louis Board of Education,* St. Louis Schools Collection, Missouri History Museum, St. Louis; letter from DuBose to Nathaniel Johnson, January 16, 1973, box 18, STLTUC.

72. Karen Van Meter, "Teachers' Choice: To Carry Textbook or Picket Sign," *SPD,* January 12, 1973.

73. "Pay Hike for Teachers 'Inadvisable,'" *SGD,* January 30, 1973. Local tax revenues accounted for just over $46 million, and close to $10 million came from the federal government.

74. Robert Kelly, "School Strike Ends," *SPD,* February 19, 1973.

75. "Bans Teacher Strike, Defiance Predicted," *SPD,* January 31, 1973; Robert Kelly, "School Strike Ends," *SPD,* February 19, 1973.

76. "No Strike Agitators Needed Here," *SGD,* January 19, 1973.

77. Ellen Sherberg and Karen Marshall, "All City Schools Closed by Board," *SGD,* January 24, 1973; "Lawless Teachers Hit New Low," *SGD,* January 25, 1973. The charge of affiliation with the Teamsters was ridiculous: the AFL-CIO expelled the International Brotherhood of Teamsters (IBT) in 1957.

78. KMOX editorial, January 25, 1973 (aired at 8:50 a.m. and 5:20 p.m.), box 18, STLTUC.

79. "Finding a Strike Solution," *SPD,* February 1, 1973; "Teachers Not above the Law," *SGD,* February 1, 1973.

80. "Teachers' 'Situation Ethics' Harm Pupils," *SGD,* February 9, 1973.

81. Gordon, *Mapping Decline,* 13–22.

82. U.S. Census, http://www.census.gov/population/www/documentation/ twps0027/twps0027.html, accessed July 15, 2016; Missouri Advocacy Committee to the U.S. Commission on Civil Rights, "School Desegregation in the St. Louis and Kansas City Areas," January 1981, 2–3; Gordon, *Mapping Decline,* 10–11.

83. "City Short-Changed on School Aid," *SGD,* January 27–28, 1973.

84. "End Illegal School Strike," *SPD,* January 23, 1973.

85. "Tough Job," *SGD,* January 29, 1973.

86. Letter to the editor from Nancy Duncan, *SPD,* February 7, 1973.

87. "Unequal Schools," *SPD,* February 11, 1973.

88. "At the Root, Money," *SPD,* February 15, 1973.

89. "What about the Governor?" *SPD,* February 16, 1973.

90. Editorial, *SGD,* May 10, 1973.

91. Mirel, *Rise and Fall of an Urban School System,* 314–15; 362–65.

92. Ibid., 362–65; quotation is on p. 363. On the accountability issue, see also David Cooper, "Accountability: Hang-Up in the Strike," *Detroit Free Press (DFP),* September 20, 1973.

93. William Grant, "Teachers Defy Work Order as Court Reopens Schools," *DFP,* September 26, 1973.

94. "Schools: Wisdom and Foolishness," *DFP,* September 26, 1973.

95. "Teachers' Strike," *DFP*, October 5, 1973.

96. "Ms. Riordan Seen as Deserving Jail," *Detroit News*, October 14, 1973.

97. "School Strikes Should Be Top Legislative Priority," *DFP*, October 4, 1973.

CHAPTER 4. DROPPING DEAD

1. "Transcript of President's Talk on City Crisis, Questions Asked and His Response," *NYT*, October 30, 1975.

2. Freeman, *Working-Class New York*, 256–290; quotation is on p. 272. See also Ferretti, *Year the Big Apple Went Bust*; Lichten, *Class, Power, and Austerity*; and Moody, *From Welfare State to Real Estate*, esp. pp. 9–61.

3. "Summary of Issues in Teachers' Contract Dispute," *Baltimore Sun*, February 4, 1974; PSTA strike flyer, box 139, Teacher's Strike file, William Donald Schaefer Papers, Baltimore City Archives, Baltimore, MD (hereafter, WDS Papers).

4. See "Teacher Demands on Empty Purse," *Baltimore Sun*, February 5, 1974; and "The City's Children," *Baltimore Sun*, February 17, 1974.

5. Mayor's Office News Release, February 21, 1974, box 139, Teacher's Strike file, WDS Papers. Jane Berger argues that Schaefer's tenure heralded the triumph of austerity politics in Baltimore, allowing the city to avoid the more substantial deficits of its neighbors New York and Philadelphia, and his stance regarding pay raises for teachers certainly fits into this characterization. See Berger, "'There is tragedy on both sides of the layoffs.'" Still, Schaefer did try to convince the State of Maryland to help Baltimore. On February 11, for instance, he provided area legislators with an articulate description of the region's political economy, highlighting how the city had to pay disproportionately for police protection and sanitation services, which were necessary for suburban residents working in the city. See "Presentation by William Donald Schaefer to Baltimore City Members of General Assembly, Feb. 11, 1974," box 139, Teacher's Strike file, WDS Papers.

6. Various estimates place the number of striking police officers from anywhere from 600 to 1,200 of the city's 2,400 police force; see Fred Barbash, "Baltimore Work Strike Ends; Police May Settle Today," *Washington Post*, July 16, 1974.

7. Letters to Mayor Schaefer, July 12, 1974 and July 8, 1974, box 190, Teachers' Strike Letters from Citizens, WDS Papers.

8. "Alioto Seeks End of School Strike," *NYT*, March 19, 1974; "School Strike Ends in San Francisco," *NYT*, March 28, 1974.

9. *NBC Nightly News*, May 2, 1974, Television News Archive, Vanderbilt University, Nashville, TN (hereafter, TNA).

10. John Kifner, "Longest Teacher Strike Divides New Hampshire Area," *NYT*, June 21, 1974; "U.S. Judge Refuses to Halt Arrest of Teacher Pickets," *NYT*, August 31, 1974; "U.S. Judge Allows Teacher Pickets," *NYT*, September 4, 1974.

11. Mertz, "1974 Hortonville Teacher Strike." Albert Shanker's "Where We Stand" column linked Timberlane and Hortonville in May 1974 in a call for AFT-NEA unity; see "Now More Than Ever, Teachers Need Unity—and Labor Support," *NYT*, May 12, 1974.

12. Bureau of Labor Statistics, "Work Stoppages Involving 1000 or More Workers, 1947–2011," http://www.bls.gov/news.release/wkstp.t01.htm, accessed July 18, 2016.

13. See, among many examples, "Walkouts Disrupting U.S. Cities," *Chicago Tribune*, July 15, 1974, and "Nearly 600 Strikes Grip Nation," *Hartford Courant*, July 17, 1974.

14. Damon Stetson, "Meany Backs Public Employee Militancy," *NYT,* Nov. 7, 1974.

15. McCartin, "A Wagner Act for Public Employees," 133–134; A. H. Raskin, "On Labor's Agenda for 1975: Millions of Public Employes," *NYT,* September 22, 1974.

16. McCartin, "'A Wagner Act for Public Employees,'" 135–36.

17. Congressional Budget Office, "New York City's Fiscal Problem: Its Origins, Potential Repercussions, and Some Alternative Policy Responses," box 21, folder 11, United Federation of Teachers Collection, Tamiment Library and Robert F. Wagner Labor Archives, Eleanor Holmes Bobst Library, New York University, New York (hereafter, UFTC); City of New York Washington Office, "New York City Fiscal Crisis: An Overview," November 20, 1975, box 070022, folder 2, Mayor Abraham Beame Collection, LaGuardia and Wagner Archives, LaGuardia Community College/ CUNY, Long Island City, Queens, New York (hereafter, MABC); Ferretti, *Year the Big Apple Went Bust,* 100.

18. Beame statement, "Hearings on the Effects of the Recession Before the Senate Subcommittee on Intergovernmental Relations" January 30, 1975, box 3, roll 26 (MN61026), Mayor Abraham Beame Papers, New York City Department of Records, Municipal Archives, New York (hereafter, MABP); Ferretti, *Year the Big Apple Went Bust,* 155; "The City's Jobless Rate Rises to 11.7%," *New York Post* (NYP), July 26, 1975; Mark Lieberman, "Standard and Poor's Drops the City's 'A' Bond Rating," *New York Daily News* (NYDN), April 3, 1975.

19. Mark Lieberman, "Standard and Poor's Drops the City's 'A' Bond Rating," *NYDN,* April 3, 1975.

20. Mark Lieberman, "City to Fire 4,000; Still Shy $641M," *NYDN,* April 23, 1975.

21. George Arzt and Steven Marcus, "New Beame Cuts: 4,000 Jobs, 43 Schools, 4 Hospitals, 10 Libraries," *NYP,* April 22, 1975; Ferretti, *Year the Big Apple Went Bust,* 187; "Simon's Surgery for City May Leave You in Stitches," *NYDN,* May 16, 1975.

22. Fred Ferretti, "Mayor Emotional: Attributes City Plight to Wall Street, Banks, Albany, Washington," *NYT,* May 30, 1975; Feretti, *Year the Big Apple Went Bust,* 199–201; "Positions Dropped in 1974/75 and 1975/76" and "Employees Laid Off: Direct Mayoral Agencies, July 25, 1975," box 070001, MABC.

23. Carl Pelleck, "Unions Await Ruling on 'Fear City' Drive," *NYP,* June 14, 1975; Carl Pelleck and Larry Kleinman, "Unions Start Distributing New 'Fear City' Leaflet," *NYP,* June 19, 1975; Ferretti, *Year the Big Apple Went Bust,* 221; "City Crimes Go Over 500,000 in 1974," *NYDN,* April 1, 1975.

24. McCartin, "Turnabout Years."

25. Helen Dudar, "Sanitmen Strike: Manhattan, Brooklyn Hit," *NYP*, June 27, 1975; "Cops Tie Up Brooklyn Bridge," *NYP*, July 1, 1975; Cy Egan and Arthur Greenspan, "A 'Total' Garbage Strike," *NYP*, July 1, 1975; Robert Crane, Vincent Lee, and Donald Singleton, "Strike Has Garbage Piling Up," *NYDN*, July 2, 1975; Patrick Doyle and Steven Matthews, "Fired Cops Wage Battle with Ins," *NYDN*, July 2, 1975; "It Gets Worse," *NYP*, July 2, 1975; Vincent Lee and Michael Patterson, "'Sick' Firemen Join in Protest," *NYDN*, July 3, 1975; Helen Dudar and Cy Egan, "Strike Ends," *NYP*, July 3, 1975; James Norman, "The Night of the Fires," *NYP*, July 3, 1975; Sam Roberts, Donald Singleton, and Mark Liff, "Rehire 2,750 Cops, Firemen: 10,000 May Be Spared," *NYDN*, July 4, 1975; Thomas Poster, "Beame to Get 330M Taxes," *NYDN*, July 4, 1975.

26. John Darnton, "M.A.C. Urges Dramatic Cuts on City to Reopen Bond Market," *NYT*, July 18, 1975.

27. Andy Soltis, "Free Tuition: A New Attack," *NYP*, July 26, 1975.

28. "Plan Rallies to Keep Free Tuition," *New York Amsterdam News* (*NYAM*), July 20, 1975.

29. "Statement by Mayor Beame," July 31, 1975, box 070012, MABC; Fred Ferretti, "Some Unions Balk," *NYT*, August 1, 1975; Ferretti, *Year the Big Apple Went Bust,* 273–89; William Sherman and Thomas Potter, "Beame Gives in 'to Save City,'" *NYDN*, September 5, 1975.

30. Frank Lombardi, "Abe Tries to Explain to Ford Midwest Has Stake in City," *NYDN*, November 3, 1975; Van Riper, "Beame Attacks Ford for 'Hate' Campaign," *NYDN*, November 6, 1975; Van Riper, "Ford Demands State Act First," *NYDN*, November 19, 1975; Mark Lieberman and Owen Moritz, "Teacher Fund and Banks Fall into Line on City Aid," *NYDN*, November 26, 1975; "Text of Ford's Statement on Help for City," *NYDN*, November 27, 1975; City of New York Washington Office, "New York City Fiscal Crisis: An Overview," November 20, 1975, box 070022, MABC; Ferretti, *Year the Big Apple Went Bust,* 366–72; Beame press release, November 11, 1975, box 070013, MABC; Francis Clines, "The President and Fiscal Politics," *NYT*, November 20, 1975; Josh Friedman, Steve Lawrence, and Steven Marcus, "Council OK's City Tax Package," *NYP*, November 22, 1975.

31. New York City Office of Management and Budget, September 30, 1975, box 07001, MABC; Freeman, *Working-Class New York,* 270–72; Leonard Buder, "Class Sizes and Rehiring Still Plague City Schools," *NYT*, October 3, 1975; "Students Boycott Classes to Protest Teacher Cuts," *NYDN*, October 4, 1975; Fred Ferretti, "Financial Crisis Crippling New York's Public Schools," *NYT*, December 12, 1976; Lewis, *New York City Public Schools,* 74; Lichten, *Class, Power, and Austerity,* 165–80.

32. Israel Shenker, "18 Urban Experts Advise, Castigate, and Console the City on Its Problems," *NYT*, July 30, 1975.

33. Congressional Budget Office, "New York City's Fiscal Problem," box 21, folder 11, UFTC.

34. Beame press release, January 30, 1975, and "Remarks by Mayor Abraham Beame at Hearings on the Effect of the Recession before the Senate Subcommittee

on Inter-governmental Relations, United States Senate," January 31, 1975, box 070012, MABC.

35. Ibid.

36. Staff of the Joint Committee on Internal Revenue Taxation, *General Explanation of the State and Local Fiscal Assistance Act of 1972* (Washington, DC: U.S. Government Printing Office, 1973), 6–7. Congress renewed the law beyond its original five-year charter, but it died, after considerable downsizing, in 1986. See James Cannon, "Federal Revenue-Sharing: Born 1972. Died 1986. R.I.P.," *NYT*, October 10, 1986.

37. Remarks by Mayor Beame, January 31, 1975, box 070012, MABC.

38. "Remarks by Mayor Abraham Beame at Public Hearings Held by the U.S. Senate Budget Committee," March 5, 1975, box 070012, MABC.

39. "Some Nerve," *NYDN,* April 7, 1975.

40. Letter from Gloria Klinga to Mayor Beame, June 11, 1975, box 7 (MN61–012), MABP.

41. Letter from O. A. Westerhaus to Mayor Beame, April 28, 1975, box 7 (MN61–012), MABP. Victor Gotbaum was president of AFSCME District Council 37 and John Delury headed the sanitation workers union.

42. Sterling and Haight, *Mass Media*, 335–36.

43. Chapman, *Tell It to Sweeney*; Squires, *Read All about It!*

44. "The Days Dwindle Down," *NYDN,* April 16, 1975.

45. "Think We Got Troubles?" *NYDN,* April 20, 1976.

46. "Something's Got to Give," *NYDN,* April 28, 1975.

47. Tifft and Jones, *Trust*, 471, 505–9.

48. " . . . And for New York City," *NYT,* April 19, 1975.

49. "'No Recourse'?," *NYT*, May 30, 1975.

50. Kosner, *It's News to Me*, 92, 113–14; Nissenson, *The Lady Upstairs*, ix, 330–49; Sterling and Haight, *Mass Media*, 335–36.

51. " . . . Casualty Lists of the Cities," *NYP*, April 15, 1975.

52. "Washington's Brushoff for New York," *NYP*, May 12, 1975.

53. "Rebuff for the City," *NYP*, May 16, 1975.

54. A. H. Raskin, "New York's Desperation Reflects That of Its Poor," *NYT,* May 18, 1975. For the biographies authored by Raskin, see A. H. Raskin, *Sidney Hillman, 1887–1946* (New York: American Jewish Committee, 1947), and David Dubinsky and A. H. Raskin, *David Dubinsky: A Life with Labor* (New York: Simon and Schuster, 1977).

55. Daily mail reports, box 7 (MN61–012), MABP.

56. Letter from Ann Volpe to Mayor Beame, January 17, 1975, box 7 (MN61–012), MABP.

57. "Undervalued Resource," *NYP,* May 9, 1975.

58. Letter from Aguilar Library Support Committee to Mayor Beame, May 7, 1975, box 7 (MN61–012), MABP.

59. "Of City Economies and Higher Education," *NYT,* May 31, 1975.

60. Letter from Agapito Otero to Mayor Beame, April 30, 1975, box 7 (MN61–012), MABP.

61. Letter from Beatrice Larkin to Mayor Beame, June 16, 1975, box 7 (MN61–012), MABP.

62. Letter from Dorothy Gunderson to Mayor Beame. No date provided, but in the June folder, box 7 (MN61–012), MABP.

63. Letter from R. W. Houseman to Mayor Beame, February (date not specified), box 7 (MN61–012), MABP.

64. "Garbage Schedule," *NYP,* July 29, 1975.

65. Nadasen, *Welfare Warriors.*

66. Cloward and Piven, "Strategy to End Poverty."

67. Quadagno, *Color of Welfare,* 120–21.

68. Reese, *Backlash against Welfare Mothers,* 113–17.

69. Letters to the editor, *NYP,* April 1, 1975.

70. "Looking for Bread," *NYDN,* May 16, 1975.

71. Letter from Marion Wertheimer to Mayor Beame, July 31, 1975, box 7 (MN61–012), MABP.

72. "The Purge," *NYP,* June 5, 1975.

73. Letter from Maureen Cullen to Mayor Beame, June 9, 1975, box 7 (MN61–012), MABP.

74. "Have Pity," *NYDN,* April 24, 1975.

75. Harriet Van Horne, "A Failing System," *NYP,* June 2, 1975.

76. Earl Graves, "The Black Stake in New York City," *NYAM,* September 17, 1975.

77. Vernon Jordan Jr., "To Be Equal," *NYAM,* July 2, 1975.

78. "Where We Stand—New York City's Fiscal Crisis: The Federal Government Must Share Blame and Cost," *NYT,* June 8, 1975.

79. Mayor Beame press release, July 31, 1975, box 070012, MABC.

80. "Strong-Arm Tactics," *NYDN,* July 2, 1975.

81. "Verbal Pollution," *NYP,* July 1, 1975.

82. "Fear Merchants," *NYP,* June 16, 1975.

83. Letter from Kendall Lutes to Mayor Beame, July 2, 1975, box 7 (MN61–012), MABP.

84. Letter from John Murray to Mayor Beame, July 9, 1975, box 7 (MN61–012), MABP.

85. John Kifner, "Over 90% of Boston Teachers Strike, Crippling Desegregation," *NYT,* September 23, 1975; "Fewer Teacher Strikes Reported, but Many Disputes Unsettled," *NYT,* September 7, 1976.

86. Seth King, "Chicago Teachers Reach an Accord," *NYT,* September 17, 1975.

87. Seth King, "24,000 Chicago Teachers Strike, Closing 666 Schools," *NYT,* September 4, 1975.

88. "Teacher Strikes Plague 11 States," *NYT,* September 5, 1975.

89. "2 Million Students across U.S. Idled by Teacher Strikes," *NYT,* September 11, 1975.

90. On the Boston desegregation controversy, see Formisano, *Boston against Busing*, and Lukas, *Common Ground*.

91. John Kifner, "Boston's Teachers Reach Tentative Strike Accord," *NYT,* September 20, 1975.

92. James Wooten, "Both Sides Stand Firm in Wilmington, Del., after Arrest of 209 Teachers in Bitter Three-Week Strike," *NYT,* September 24, 1975; "Wilmington Strike by Teachers Ended," *NYT,* October 11, 1975.

93. B. Drummond Ayres Jr., "Atlanta Teachers Strike; Return to Work Ordered," *NYT,* October 15, 1975.

94. Lawrence Fellows, "New Haven Strike Ends: Jailed Teachers Freed," *NYT,* November 25, 1975.

95. Mayor Beame press release, September 10, 1975, box 070012, MABC.

96. Warren Kung, "To Hell with the City," and "A Teachers' Strike," *NYDN,* September 10, 1975.

97. "Shanker's Fear Tactics," *NYT,* July 25, 1975.

98. "A No-Win Folly," *NYP,* September 9, 1975.

99. "The Teachers' Case," *NYT,* September 17, 1975.

100. "City in Crisis," *NYDN,* September 7, 1975.

101. "A Bum Contract," *NYDN,* September 23, 1975.

102. On responses to the fiscal crisis, see "CORE Marches to Dramatize City's Crisis," *NYAM,* June 25, 1975; "Black Teachers to Stage Protest at City Hall," *NYAM,* October 15, 1975.

103. "Mr. Shanker Takes the Floor," *NYAM,* July 20, 1975; see also "Black Schools Won't Follow Shanker in School Strike," *NYAM,* September 10, 1975.

104. "Irresponsible Brinksmanship," *NYDN,* October 18, 1975.

105. Howard K. Smith editorial comment, *ABC Evening News,* October 16, 1975, TNA.

106. Letter from John Santore to Mayor Beame, November 17, 1975, box 7 (MN61–012), MABP.

107. *NBC Nightly News,* November 30, 1975, TNA.

CHAPTER 5. THE PITTSBURGH TEACHER STRIKE OF 1975–76 AND THE CRISIS OF THE LABOR-LIBERAL COALITION

1. "Walkouts Disrupting U.S. Cities," *CT,* July 15, 1974; "Nearly 600 Strikes Grip Nation," *Hartford Courant,* July 17, 1974.

2. Irwin Ross, "The Fiscal Follies of New York City," *Reader's Digest,* October 1975.

3. Vermont Royster, "The Undoing of Great Britain: A Textbook Case of How to Ruin a Once-Vigorous Economy," *Reader's Digest,* January 1976.

4. Sterling and Haight, *Mass Media,* 346–50.

5. Sharp, *Condensing the Cold War,* 47.

6. Howard K. Smith comment, *ABC Evening News,* December 9, 1975, Television News Archive, Vanderbilt University, Nashville, TN (hereafter, TNA).

7. *ABC Evening News*, June, 5, 1975, TNA.

8. Ibid.; "PFT Demands Could Cost Taxpayers over $30 Million," *New Pittsburgh Courier (NPC)*, December 6, 1975.

9. Edward Fiske, "City Is Rated High in Teacher Pay," *NYT*, September 8, 1975.

10. James Wooten, "Walkout by Pittsburgh Teachers Brings Frustration and Impasse," *NYT*, January 17, 1976.

11. Joyce Gemperlein, "PTA Backs Suit in School Strike," *PPG*, December 17, 1975.

12. James Wooten, "Walkout by Pittsburgh Teachers . . .," *NYT*, January 17, 1976.

13. "Eight-Week School Strike Ends," *PPG*, January 27, 1976.

14. "Strike Ends," *NPC*, January 31, 1976.

15. "Mayor Pete Flaherty Address to the City Council," *PPG*, December 13, 1975.

16. Letter to the editor, *PP*, January 14, 1976.

17. Trotter and Day, *Race and Renaissance*, 120.

18. Ibid., 46, 53, 81–85; Diane Perry, "The Coming Crisis in Pittsburgh Schools," *NPC*, August 29, 1970.

19. "PFT Threatens New Strike," *NPC*, May 2, 1970; "NAACP Rejects PFT Separate School Plan," *NPC*, June 13, 1970.

20. *Report of the Select Commission to Study the Pittsburgh School Board* (1969), box 136, and "Pittsburgh Public Schools Membership Report as of Oct. 4, 1976," box 51, Pittsburgh Public Schools Collection, Heinz Historical Center Archives, Pittsburgh, PA (hereafter, HHCA).

21. Robert Flipping Jr., "Teachers Demand Disciplinary Action Be Enforced by Board," *NPC*, May 24, 1975; "Pittsburghers Respond to Proposal Regarding Paddling," *NPC*, November 1, 1975.

22. "Pittsburghers Respond to Proposal Regarding Paddling," *NPC*, November 1, 1975.

23. "Teacher Strike, Residency Requirement Draw Comments," *NPC*, November 22, 1975.

24. "Another Blow to the Schools," *PPG*, December 2, 1975.

25. Letters to the editor; "Shouldn't Schoolteachers Be Paddled If They Walk Out on a Strike?," *PPG*, November 29, 1975.

26. "Rep. Al Fondy?," *PPG*, December 3, 1975.

27. "A Living Wage," *PPG*, December 4, 1975.

28. "A Parent's Answer to School Strike," *PPG*, December 6, 1975.

29. "Money for Teachers," *PPG*, December 8, 1975.

30. "Time to Teach!" *PPG*, December 10, 1975; "My Solution" and "Clarion Call," *PPG*, December 12, 1975.

31. "A Teacher's Reply," *PPG*, December 9, 1975.

32. "City School Strike Hurts Blacks," *PPG*, December 15, 1975.

33. "Just Raise Taxes?," *PPG*, December 16, 1975.

34. "No Strike Right," *PPG*, December 30, 1975.

35. "Calls Teachers Greedy," *PPG*, December 31, 1975.

36. "Judge Orders Teachers Back in Class Tomorrow; Will Defy Injunction, Fondy Says," *PP*, January 4, 1976.

37. "Teachers Should Obey," *PPG*, January 6, 1976.

38. "Defying the Law," *PP*, January 6, 1976.

39. Cy Hungerford, "The Spirit of '76," and "Twisted Points," *PPG*, January 8, 1976. The suburban *North Hills News Record* weekend edition of January 10–11 also published Beyer's letter.

40. Letters to the editor, *PP*, January 9, 1976.

41. "Teachers' Union Tactics Denounced," "What Fondy Teachers Youth," "Teachers Set Bad Example," *PP*, January 9, 1976; McCartin, "'Fire the Hell Out of Them.'"

42. "Teacher Backer Changes Mind," *PP*, January 11, 1976.

43. Letter to the editor, *PP*, January 10, 1976. For an insightful treatment of memory and steelworkers, especially with regard to the 1959 strike, see Metzgar, *Striking Steel*.

44. "People's Forum," *Valley Independent* (Monessen, PA), January 10, 1976.

45. On the 1970s and moral crisis, see Christopher Lasch's *A Culture of Narcissism: American Life in an Age of Diminishing Expectations* (New York: W. W. Norton, 1978), as well as historian William Graebner's chapter "America's *Poseidon Adventure*: A Nation in Existential Despair," in *America in the Seventies*, ed. Beth Bailey and David Farber (Lawrence: University of Kansas Press, 2004), 157–80.

46. Cy Hungerford, "The Strikers Kept Walking," and "Fondy Recollection," *PPG*, January 12, 1976.

47. "Productivity and Stability," *PPG*, January 12, 1976.

48. "Teachers: The New Priesthood," *PPG*, January 14, 1976.

49. See, for instance, Eugene Methvin's "The NEA: A Washington Lobby Run Rampant," *Reader's Digest*, November 1978, 97–101. For more recent vintage, see Myron Lieberman's *The Teacher Unions: How the NEA and the AFT Sabotage Reform and Hold Students, Parents, Teachers, and Taxpayers Hostage to Bureaucracy*.

50. "Intolerable Strike," "Taxpayers, Do It!," *PPG*, January 15, 1976.

51. "Crackdown Urged on PFT," *PP*, January 16, 1976.

52. "Let the Taxpayers Have a Say, Too," *North Hills News-Record*, January 24–25, 1976.

53. "'No More Leftovers' for Teachers," *PP*, January 16, 1976.

54. "The Strike Debate Goes On and On," *PP*, January 21, 1976.

55. Mason Denison, "Teacher Strikes Underscore Growing Public Helplessness," *Kitanning Leader Times*, January 21, 1976.

56. "Strike Lessons," *PP*, January 29, 1976.

57. Cowie, *Stayin' Alive*, chap. 3.

58. Ira Fine, "Labor's Outlook 'Boggles Mind,'" *PP*, January 13, 1976.

59. See Jon Shelton, "'Against the Public': Teacher Strikes and the Decline of Labor-Liberalism, 1968–81" (PhD diss., University of Maryland, College Park, 2013), chap. 5. In my dissertation, I show that the NRTWC argued that the Detroit Federation of Teachers violated both its members and the public by signing an "agency fee" arrangement with the Detroit school board in 1969. Ultimately, the NRTWC backed a lawsuit that eventually reached the Supreme Court: *Abood v. Detroit Board of*

Education (1977). The Supreme Court upheld agency fees, and this case still represents the most important precedent regarding the agency shop and public-employee unions in the United States.

60. James Scott testimony to Governor's Study Commission, Harrisburg, April 5, 1977, RG-16, carton 1, Governor's Commission for the Revision of the State's Public Employee Law Records, Pennsylvania State Archives, Harrisburg, PA (hereafter, PSA).

61. W. Thacher Longstreth testimony to Governor's Study Commission, Pittsburgh, April 26, 1977, RG-16, carton 3, PSA.

62. Melvin Miller and Jerry Olsen testimony, Governor's Study Commission, Pittsburgh, May 12, 1977, RG-16, carton 3, PSA.

63. Edna Irving testimony, Governor's Study Commission, Philadelphia, April 26, 1977, RG-16, carton 3, PSA.

64. Bob McCannon testimony, Governor's Study Commission, Philadelphia, April 27, 1977, RG-16, carton 3, PSA.

65. See *Recommendations for Legislative and Administrative Change to the Public Sector Collective Bargaining Laws of Pennsylvania* (Harrisburg, PA: Governor's Study Commission, 1978).

66. Albert Fondy testimony, Governor's Study Commission, Harrisburg, April 5, 1977, RG-16, carton 1, PSA.

CHAPTER 6. THE "FED-UP TAXPAYER"

1. Johnston, *Success While Others Fail*, 4.

2. "Fewer Teacher Strikes Reported but Many Disputes Unsettled," *NYT*, September 7, 1976.

3. "Teachers Vote to End 11-Day Oakland Strike," *LAT*, November 16, 1977.

4. Ronald Smothers, "Lakeland Teachers Accept Pact, Ending State's Longest School Strike," *NYT*, November 10, 1977.

5. "Teachers in Toledo End Strike That Shut Schools for 23 Days," *NYT*, May 3, 1978.

6. Gregory Jaynes, "Wilmington Teachers End Strike with Approval of Three-Year Pact," *NYT*, November 22, 1978.

7. Patricia Cullinen, "Why My Husband Went to Jail," *NYT*, December 10, 1978.

8. Alfonso Narvaez, "Judge in Newark Averts Walkout Set by Teachers," *NYT*, January 12, 1979.

9. Kuttner, *Revolt of the Haves*, 92.

10. Ronald Reagan, "Tax Cuts Promise Economic Growth," *SGD*, June 14, 1978.

11. See, for instance, James Roper, "California Tax Vote Sends Shock Waves," *PP*, June 4, 1978; Louise Cook, "California Vote Expected to Fuel Taxpayer Revolt," *SGD*, June 8, 1978.

12. Eric Mink, "53-Day Strike Is Over, Papers Are Back," *SPD*, January 14, 1979; "Millions Face Layoffs in Britain as Result of Work Stoppages," *SPD*, January 15, 1979.

13. Thomas Ottenad, "President's Austerity Budget Augurs Bitter Struggle," *SPD*, January 22, 1979.

14. This figure includes about 4,000 teachers and 400 additional employees.

15. James Ellis, "St. Louis Teachers Reject Wage Offer, Vote to Strike," *SPD*, January 15, 1979. There is ample evidence that rank-and-file teachers supported such an enthusiastic response. See, for example, Jo Mannes, "Teachers in High Spirits, Promise to Support Strike," *SPD*, January 17, 1979, and Linda Eardley, "400 Teachers March in Solidarity," *SPD*, January 26, 1979, which recounts a march of teachers singing "We Shall Overcome" from the Gateway Arch to the board of education building downtown.

16. James Ellis, "Teachers' Strike Boils Down to One Issue: Money," *SPD*, January 21, 1979, and "St. Louis Teachers Reject Wage Offer, Vote to Strike," *SPD*, January 15, 1979. The *St. Louis Post-Dispatch*, hardly an advocate for the striking teachers, pointed out that in 1977–78, the board underestimated its revenues by $5 million. See "A Strike against History," February 18, 1979.

17. "Board of Education of the City of St. Louis vs. Evelyn Battle, President, et al., St. Louis Teachers Union," January 15, 1979, box 18, St. Louis Teachers Union Local 420 Records, box 18, Walter P. Reuther Library, Wayne State University, Detroit, MI (hereafter, STLTUC); James Ellis, "Teachers Defy Order on Strike," *SPD*, January 16, 1979; Geoff Dubson, "Union Tells Teachers to Expect No Mercy from Judges," *SPD*, January 22, 1979.

18. St. Louis Labor Council press release, January 26, 1979, box 18, STLTUC.

19. James Ellis, "Board Drops Suit against Teachers," *SPD*, January 29, 1979.

20. James Ellis, "Teasdale to Meet with Teachers, School Board," *SPD*, February 24, 1979.

21. James Ellis, "Teasdale Sees 'Ray of Hope' in $1.4 Million School Offer," *SPD*, March 3, 1979; Jay Pulitzer and James Ellis, "Deputies Barge in on Teacher Talks with Order to End Strike," *SPD*, March 10, 1979.

22. James Ellis and Jay Pulitzer, "Union Head Urges Teachers to Ratify Pact," *SPD*, March 12, 1979.

23. "Teach Again, Talk Again," *SPD*, January 17, 1979.

24. "Californians Kill 'the Monster,'" *SGD*, June 8, 1978; Don Herse, "It's the Aftermath from the California Earthquake," *SGD*, June 10–11, 1978.

25. "Striking Teachers Flunk Test," *SGD*, January 17, 1979.

26. "Supports Teachers' Strike," *SGD*, January 22, 1979.

27. "Letters on the Teachers' Strike," *SPD*, February 9, 1979.

28. "The Teachers' Turn," *SPD*, January 30, 1979.

29. "School Strike Questions," *SGD*, February 1, 1979.

30. "City Teachers' Strike," *SPD*, February 15, 1979.

31. "Strike Costs," *SPD*, February 8, 1979.

32. "More Letters on the School Strike," *SPD*, February 23, 1979.

33. Ibid.

34. "More on the City's Striking Teachers," *SPD*, February 28, 1979.

35. "Two Sides," *SGD*, March 9, 1979.

36. "The School Strike: How Much Do Parents Care?," *SPD*, March 7, 1979.

37. "Not in Six Hours," *SPD*, March 8, 1979.

38. "Schools and Strikes," *SPD*, March 9, 1979.

39. "Outraged," *SGD*, March 16, 1979; "More Backlash against Teachers' Strike," *SGD*, March 17–18, 1979.

40. "Washington Teachers End a Strike after Court Reinstates Labor Pact," *NYT*, March 29, 1979.

41. "Thousands of Teachers Strike in Parts of Eight States," *NYT*, August 28, 1979.

42. "Over 664,000 Students Affected by Teachers' Strikes in 11 States," *NYT*, September 25, 1979.

43. William Stevens, "Cleveland's Schools, Failing to Meet Payroll, Enter Deeper Crisis," *NYT*, November 25, 1977.

44. Iver Peterson, "Defeat of School Levy in Cleveland Strikes Angry Blow at Busing Plan," *NYT*, April 16. 1978.

45. "Teachers' Walkout in Cleveland Shuts Down Most of the School System," *NYT*, October 19, 1979; "Cleveland Teachers Vote to End 11-Week Strike," *NYT*, January 4, 1980.

46. Bureau of Labor Statistics, "Unemployment Rate from 1947 to 2013," http://data.bls.gov/timeseries/LNU04000000?years_option=all_years&periods _option=specific_periods&periods=Annual+Data, accessed July 19, 2016.

47. Stephen Williams, "Board Ready to Discuss Closing of 15 Schools," *PT*, April 11, 1980; Steve Twomey, "School Strike: A Quiet First Day," *PI*, September 2, 1980.

48. Ray Holton and Steve Twomey, "School Talks Resume," *PI*, September 3, 1980; Steve Twomey, "Stalemate on a Principle Means School-Strike Standoff," *PI*, September 13, 1980. On the rank and file, see, for instance, the comment of Thaddeus Jones, a striking teacher, who put it as follows: "It's enough to disillusion us. After we go to four years of college, you look forward to some kind of security in the teaching profession"; Jan Pogue, "On the Teachers' Picket Line," *PI*, September 3, 1980.

49. Ray Holton and Steve Twomey, "Why the Plan to End the School Strike Died," *PI*, September 6, 1980; "Mayor's Statement," *PI*, September 9, 1980.

50. McCartin, "'Fire the Hell Out of Them.'"

51. "Teachers and $$$," *Philadelphia Bulletin* (*PB*), September 24, 1980.

52. McCartin, *Collision Course*.

53. United States Department of Education, *Nation at Risk*, 1.

54. "Public Attitudes toward Television and Other Media in a Time of Change: The Fourteenth Report in a Series by the Roper Organization, Inc." (New York: Television Information Office, 1985), 13.

55. Mary Bishop, Thomas Ferrick Jr., and Donald Kimelman, "The Real Power in the District: The Teachers," *PI*, September 1, 1981; Mary Bishop, Thomas Ferrick Jr., and Donald Kimelman, "How Inept Teachers Thrive in the District," *PI*, September 1, 1981.

56. Dan Freedman, "Have Union Pacts for Teachers Gotten Out of Control?," *PB*, September 14, 1981.

57. Thomas Ferrick Jr, "Green Wants a 10% Tax Hike," *PI*, September 1, 1981.

58. Moody, *Injury to All*, 165–92.

59. "Excerpts from the Mayor's Speech on the Schools," *PI*, September 1, 1981.

60. Rick Nichols, "High-Level Talks Held on Schools," *PI*, September 3, 1981; Dan Freedman, "City Schools and Union Agree: Shift in U.S. Aid a 'Disaster,'" *PB*, September 2, 1981.

61. Rick Nichols and Mary Bishop, "Opening of Schools in Doubt," *PI*, September 9, 1981; Maurice Lewis Jr. and Craig McCoy, "Phila. Teachers Strike: Pickets Posted, Bitterness Seen," *PI*, September 6, 1981.

62. "206 Teachers Seized," *PB*, September 9, 1981.

63. Mary Bishop and Terry Johnson, "A School Opens, Bitterly," *PI*, September 19, 1981; Craig McCoy and Elmer Smith, "End Strike: Schools Take Teachers to Court," *PB*, September 17, 1981; Craig McCoy, "Phila. School Chief's Stand on Strike Comes as Shock to Teachers' Union," *PB*, September 20, 1981; Kevin Goldman, Joyce Gemperlein, and Marc Schogol, "Most Seniors Register: 11,600 Make Way through Strikers," *PI*, September 24, 1981.

64. Mary Bishop, "Black Groups Push to End School Strike," *PI*, October 1, 1981; open letter to William Green from Joseph Watlington, Jr., et al., October 26, 1981, RG60-2.7, box A-5576, Mayor William Green Papers, 1981, City of Philadelphia, Department of Records, City Archives (hereafter, MWGP).

65. Joyce Gemperlein, Lucinda Fleeson, and Mary Bishop, "Court Orders Teachers Back," *PI*, October 8, 1981; Craig McCoy, "Teachers Ordered to Work," *PB*, October 8, 1981.

66. Mary Bishop and Thomas Ferrick Jr., "Teachers Mass at N. Phila. School," *PI*, October 10, 1981.

67. Stephen Salisbury and Mary Bishop, "Stay Out, Murray Tells Teachers," *PI*, October 12, 1981; Joyce Gemperlein, Lucinda Fleeson, and Mary Bishop, "PFT Found Guilty of Contempt," *PI*, October 14, 1981.

68. Mary Bishop, Lucinda Fleeson, and Joyce Gemperlein, "Citywide Strike Urged for PFT," *PI,* October 15, 1981.

69. See, for instance, *CBS Evening News*, October 13, 1981, Television News Archive, Vanderbilt University, Nashville, TN.

70. Rubio, *There's Always Work at the Post Office*, chap. 10–11; McCartin, *Collision Course*, chap. 4–6.

71. Lucinda Fleeson, "Phila. Labor in Step with the March," *PI*, September 20, 1981; "Labor Sees Public Ire as Answer," *PB*, September 21, 1981.

72. Lucinda Fleeson, Linda Loyd, and Thomas Ferrick Jr., "Union Heads Vote to Back Mass Strike," *PI*, October 27, 1981.

73. Lucinda Fleeson, "Labor Hesitates at PFT's Call to Shut the City," *PI*, October 18, 1981.

74. Ibid.

75. Halle, *America's Working Man*, 48–50.

76. "Excerpts from the Court Ruling That Resulted in the End of the Strike," *PI*, October 29, 1981.

77. "A Demand for School Tax Rebate," *PB*, September 21, 1980.

78. "Mr. Green Must Insist on Sound School Contract," *PI*, September 3, 1980.

79. "Teachers Should Face Reality on Layoffs Issue," *PI*, September 8, 1980.

80. "Waiting Out the Teachers," *PB,* September 16, 1980; italics in original.

81. "School Strike: A Way Out," *PB*, September 22, 1980.

82. Acel Moore, "Both Sides Are Ignoring the Bottom Line," *PI*, September 4, 1980.

83. "Speak Now," *PI*, September 19, 1980.

84. "Not Convinced on Teacher Prep Time," *PI*, September 7, 1980.

85. "Ignored Facts," *PI*, September 19, 1980.

86. "Reactions to the Teachers' Strike," *PI*, September 11, 1980.

87. "No Savings in Firing Teachers," *PB*, September 14, 1980.

88. Letter from Harold Sorgenti to Green, August 28, 1981, box A-5576, MWGP.

89. "Is the School Dispute Politics—Or Survival?" *PI*, September 3, 1981.

90. "The Crisis in Philadelphia's Schools" and "Hang Tough," *PI*, September 6, 1981.

91. Letter from C. J. Swartz to Mayor Green, September 2, 1981, box A-5576, MWGP.

92. "No Tax Hike," *PI*, September 8, 1981; "Open the Schools," *PB*, September 11, 1981.

93. Letter from Carol Aff to Mayor Green, September 2, 1981, box A-5576, MWGP.

94. "Time for Action," *PI*, September 19, 1981.

95. Letter to Mayor Green, name illegible, September 15, 1981, box A-5576, MWGP.

96. "Willing to Pay," *PI*, October 4, 1981.

97. Letter from [name redacted by the archive] to Mayor Green, October 20, 1981, box A-5576, 1981, MWGP. Green returned the check.

98. John Murray, "Union Position: Teachers Don't Want a Strike," *PI*, September 6, 1981; "Why Go Back on the Contract?" *PI*, September 3, 1981; "Anti-Labor," *PI*, September 10, 1981.

99. "A Precedent," *PI*, September 10, 1981; "Meaning," *PI*, October 1, 1981.

100. "Not Convinced on Teacher Prep Time," *PI*, September 7, 1980; "Teacher Prep," *PI*, September 10, 1981.

101. "Striking Teachers Have Become 'Greedy,'" *PB*, September 10, 1981; "Take a Look," *PI*, September 21, 1981.

102. Dan Rottenberg, "Can Philadelphians Get Along without Their Schools," *PI*, September 29, 1981; "School Questions," *PI*, October 12, 1981; "A New System," *PB*, September 19, 1981.

103. Letter from Edna Williams to Mayor Green, September 21, 1981, box A-5576, MWGP.

104. Letter from G. Patrick Armstrong to Mayor Green, September 22, 1981, box A-5576, MWGP.

105. Russell and Rosalind Jackson, "The 'Rotten Plum' of Selfishness," *PB*, September 27, 1981.

106. "Charge Tuition," *PB*, September 26, 1981; "Disgust over Strike," *PB*, October 18, 1981.

107. Milton Friedman, "The Role of Government in Education," http://faculty.smu.edu/Millimet/classes/eco4361/readings/friedman%201955.pdf, accessed July 19, 2016; Jim Carl, *Freedom of Choice*, chap. 1–3.

108. Rusher, *Making of the New Majority Party*, 130–31.

109. Lanny Ebenstein, *Milton Friedman*, 200–203.

110. *Free to Choose*, episode 6, "What's Wrong with Our Schools?," directed by Peter Robinson (Video Arts TV Production, 1980).

111. Ibid.

112. Coons and Sugarman, *Education by Choice*.

113. *Free to Choose*, episode 6, "What's Wrong with Our Schools?," directed by Peter Robinson (Video Arts TV Production, 1980).

114. Dan Rottenberg, "The Schools Are Asked the Wrong Questions," *PI*, September 23, 1980.

115. "Time for a School Compromise," *PB*, September 2, 1981. The *Bulletin* referenced the potential of a voucher system again just five days later; see "Labor Day, 1981," *PB*, September 7, 1981.

116. Neil Peirce, " . . . But There Are Alternatives to the Present System," *PI*, October 5, 1981. See also Neil Peirce, "It Is Time to Give School Vouchers a Try . . ." *PI*, October 19, 1981.

117. "Voucher System," *PI*, October 29, 1981; "Voucher System," *PB,* September 26, 1981; "Competition," *PB*, September 12, 1981.

118. "Destruction," *PI*, October 6, 1981.

119. Ronald Reagan, "Inaugural Address, January 20, 1981," Ronald Reagan Presidential Library, National Archives and Records Administration, https://reaganlibrary.archives.gov/archives/speeches/1981/12081a.htm, accessed July 19, 2016.

CONCLUSION

1. Casey Banas and Michele Norris, "Teacher Strike in Record 16th Day," *CT,* September 29, 1987; "1983, 1984, 1987: In Teachers' Strikes, Chicago Students Suffer," *CT,* July 20, 2012.

2. Donald Perkins, "Chicago's Public School System Must Be Restructured," *CT,* September 16, 1987. On the century-long effort by the Commercial Club to influence Chicago schools along corporate lines, see Shipps, *School Reform, Corporate Style*.

3. Joetta Sack, "Teacher Strikes Limited to a Few Small Districts," *Education Week,* November 8, 2005; Elizabeth Weaver, "Pennsylvania Teachers: Number One in Strikes," *Allegheny Institute for Public Policy Report* #07–06, August 2006, 2–3, http://www.alleghenyinstitute.org/wp-content/uploads/components/com_reports/uploads/07_06.pdf, accessed July 19, 2016.

4. Google Books NGram Viewer search for "teacher strike, teacher union," https://

books.google.com/ngrams/graph?content=teacher+strike%2C+teacher+union
&year_start=1900&year_end=2007&corpus=15&smoothing=3&share=&direct
_url=t1%3B%2Cteacher%20strike%3B%2Cc0%3B.t1%3B%2Cteacher%20union
%3B%2Cc0, accessed July 19, 2016.

5. Lieberman, *Teacher Unions*; Paige, *War against Hope*; Moe, *Special Interest*.

6. Jacob Hacker and Paul Pierson, "The Wisconsin Fight Isn't about Benefits. It's about Labor's Influence," *Washington Post*, March 6, 2011; Patrick Healy and Monica Davey, "Behind Scott Walker, a Longstanding Conservative Alliance against Unions," *NYT*, June 8, 2015.

7. "Transcript of Chris Christie's Speech at the Republican National Convention," http://www.foxnews.com/politics/2012/08/28/transcript-chris-christie-speech-at -republican-national-convention/#ixzz2JxYyfOkE, accessed July 19, 2016.

8. "Christie: Teachers Union Deserves 'Punch in the Face,'" CNN.com, August 2, 2015, http://www.cnn.com/videos/politics/2015/08/02/sotu-tapper-christie-national -teachers-union-deserves-a-punch-in-the-face.cnn, accessed July 19, 2016.

9. Women increased from about 65 percent of the teaching force in 1971 to just above 70 percent by the early 1990s. In the last available set of statistics (2011–12), women represented 76.3 percent. See "Table 68: Selected Characteristics of Public School Teachers: Spring 1961 to Spring 1991," National Center for Education Statistics, *Digest of Education Statistics,* 2013, http://nces.ed.gov/programs/digest/d13/ tables/dt13_209.10.asp, accessed July 19, 2016.

10. On Rhee, see, for instance, Amanda Ripley, "Rhee Tackles Classroom Challenge," *Time,* November 26, 2008.

11. Ravitch, *Reign of Error*, esp. chap. 1–4.

12. Dan McQuade, "Bill That Approved Cigarette Tax Guarantees More Charter Schools in Philadelphia," *Philadelphia Magazine,* September 14, 2014, http://www .phillymag.com/news/2014/09/29/bill-approved-cigarette-tax-guarantees-charter -schools-philadelphia/, accessed July 19, 2016.

13. Kelly Bayliss, "School Reform Commission Cancels Contract with Philadelphia Federation of Teachers," NBC10.com, October 6, 2014, http://www.nbcphiladelphia.com/ news/politics/School-Reform-Philadelphia-SRC-278221631.html, accessed July 19, 2016.

14. Piketty, *Capital in the Twenty-First Century*.

15. Rousmaniere, *Citizen Teacher,* chap. 2–3.

16. On the CTU strike, see Alter, "'It Felt Like Community,'" and Uetricht, *Strike for America*.

BIBLIOGRAPHY

Primary Sources

Major Archival Collections

AFL-CIO Collection, George Meany Archives, University of Maryland, College Park, MD

American Federation of Teachers President's Office Records, Walter P. Reuther Library, Wayne State University, Detroit, MI (AFTPOR)

Governor's Commission for the Revision of the State's Public Employee Law Records, Pennsylvania State Archives, Harrisburg, PA (PSA)

Mayor Abraham Beame Papers, New York City Department of Records, Municipal Archives, New York (MABP)

Mayor Abraham Beame Collection, LaGuardia and Wagner Archives, LaGuardia Community College/CUNY, Long Island City, Queens, NY (MABC)

Mayor William Green Papers, City of Philadelphia, Department of Records, City Archives, Philadelphia, PA (MWGP)

Philadelphia Federation of Teachers Collection, Urban Archives, Temple University Library, Philadelphia, PA (PFTC/UA)

Pittsburgh Public Schools Collection, Heinz Historical Center Archives, Pittsburgh, PA (HHCA)

St. Louis Schools Collection, Missouri History Museum, St. Louis, MO

St. Louis Teachers Union Local 420 Records, Walter P. Reuther Library, Wayne State University, Detroit, MI (STLTUC)

Television News Archive, Vanderbilt University, Nashville, TN (TNA)

United Federation of Teachers Collection, Tamiment Library and Robert F. Wagner Labor Archives, Eleanor Holmes Bobst Library, New York University, New York (UFTC)

William Donald Schaefer Papers, Baltimore City Archives, Baltimore, MD (WDS Papers)

Major Newspapers

Baltimore Sun
Boston Globe
Chicago Daily Defender (CDD)
Chicago Tribune (CT)
Detroit Free Press (DFP)
Detroit News
Harrisburg Patriot
Hartford Courant
Los Angeles Times (LAT)
New Pittsburgh Courier (NPC)
New York Amsterdam News (NYAM)
New York Daily News (NYDN)
New York Post (NYP)
New York Times (NYT)
Newark Star-Ledger (S-L)
Philadelphia Bulletin (PB)
Philadelphia Evening Bulletin (PEB)
Philadelphia Inquirer (PI)
Philadelphia Tribune (PT)
Pittsburgh Post-Gazette (PPG)
Pittsburgh Press (PP)
St. Louis Globe-Democrat (SGD)
St. Louis Post-Dispatch (SPD)
Washington Post

Published Primary Sources

Alcaly, Roger, and David Mermelstein, eds. *The Fiscal Crisis of American Cities: Essays on the Political Economy of Urban America with Special Reference to New York.* New York: Vintage Books, 1976.

Braun, Robert J. *Teachers and Power: The Story of the American Federation of Teachers.* New York: Touchstone, 1973.

Chapman, John. *Tell It to Sweeney: The Informal History of the "New York Daily News."* New York: Doubleday, 1961.

Coons, John, and Stephen Sugarman. *Education by Choice: The Case for Family Control,* 2nd ed. Troy, NY: Educator's International Press, 1999.

De Toledano, Ralph. *Let Our Cities Burn.* New Rochelle, NY: Arlington House, 1975.

Ferretti, Fred. *The Year the Big Apple Went Bust.* New York: Putnam, 1976.

Friedman, Harvey. "The Role of the AFL-CIO in the Growth of Public Sector Collective Bargaining." In *Sorry . . . No Government Today: Unions vs. City Hall,* edited by Robert Walsh, 27–33. Boston: Beacon Press, 1969.

Friedman, Milton. *Capitalism and Freedom,* 40th anniversary ed. (Chicago: University of Chicago Press, 2002).

Friedman, Milton, and Rose Friedman. *Free to Choose: A Personal Statement.* New York: Harcourt Brace Jovanovich, 1980.

Friedman, Milton, and Anna Jacobson Schwartz, *A Monetary History of the United States, 1867–1960.* Princeton, NJ: Princeton University Press, 1971.

Galbraith, John Kenneth. *The Affluent Society.* Boston: Houghton Mifflin, 1958.

The Gallup Poll: Public Opinion, 1935–1971. Vol. 3, *1959–1971.* New York: Random House, 1972.

Journal of the Board of Education of Philadelphia, 1972–73.

Mills, C. Wright. *The New Men of Power: America's Labor Leaders.* New York: Harcourt, Brace, 1948.

———. *The Power Elite.* New York: Oxford University Press, 1956.

———. *White Collar: The American Middle Classes.* 1951. Repr., 50th anniversary ed., New York: Oxford University Press, 2002.

Orfield, Gary. *Must We Bus? Segregated Schools and National Policy.* Washington, DC: Brookings Institution, 1978.

Piven, Frances Fox. "Militant Civil Servants in New York City." *Society* 7 (1969): 24–28.

Public Attitudes toward Television and Other Media in a Time of Change: The Fourteenth Report in a Series by the Roper Organization, Inc. New York: Television Information Office, 1985.

Recommendations for Legislative and Administrative Change to the Public Sector Collective Bargaining Laws of Pennsylvania. Harrisburg, PA: Governor's Study Commission, 1978.

Resnik, Harry. *Turning on the System: War in the Philadelphia Public Schools.* New York: Pantheon Books, 1970.

Rusher, William. *The Making of the New Majority Party.* Lanham, MD: Sheed and Ward, 1975.

Selden, David. *The Teacher Rebellion.* Washington, DC: Howard University Press, 1985.

Simon, William E. *A Time for Truth: A Distinguished Conservative Dissects the Economic and Political Policies That Threaten Our Liberty—and Points the Way to an American Renaissance.* New York: Reader's Digest Press, 1978.

Ture, Kwame, and Charles Hamilton. *Black Power: The Politics of Liberation.* 1967. Repr. New York: Vintage, 1992.

United States Department of Education. *A Nation at Risk: The Imperative for Educational Reform.* 1983.

United States Department of Labor. *State Profiles: Current Status of Public Sector Labor Relations.* 1971.

Whyte, William H. *The Organization Man.* New York: Simon and Shuster, 1956.

Secondary Sources

Alter, Tom. "'It Felt Like Community': Social Movement Unionism and the Chicago Teachers Union Strike of 2012." *Labor: Studies in Working-Class History of the Americas* 10 (Fall 2013): 11–25.

Battista, Andrew. *The Revival of Labor Liberalism: The Labor-Liberal Alliance in Late Twentieth-Century American Politics*. Urbana: University of Illinois Press, 2008.

Berger, Jane. "'There is tragedy on both sides of the layoffs': Privatization and the Urban Crisis in Baltimore." *International Labor and Working-Class History* 71 (Spring 2007): 29–49.

Biondi, Martha. *To Stand and to Fight: The Struggle for Civil Rights in Postwar New York City*. Cambridge, MA: Harvard University Press, 2003.

Blumenfeld, Samuel. *NEA: Trojan Horse in American Education*. Boise, ID: Paradigm, 1984.

Boris, Eileen, and Jennifer Klein. *Caring for America: Home Care Workers in the Shadow of the Welfare State*. New York: Oxford University Press, 2012.

Brenner, Aaron, Robert Brenner, and Cal Winslow, eds. *Rebel Rank and File: Labor Militancy and Revolt from Below in the Long 1970s*. London: Verso, 2010.

Brimelow, Peter. *The Worm in the Apple: How the Teacher Unions Are Destroying American Education*. New York: Harper Collins, 2003.

Burkholder, Zoë. "'A War of Ideas': The Rise of Conservative Teachers in Wartime New York City, 1938–1946." *History of Education Quarterly* 55 (2015): 218–43.

Burns, Jennifer. *Goddess of the Market: Ayn Rand and the American Right*. New York: Oxford University Press, 2009.

Carl, Jim. *Freedom of Choice: Vouchers in American Education*. Santa Barbara, CA: Praeger, 2011.

Carter, Dan. *The Politics of Rage: The Origins of the New Conservatism, and the Transformation of American Politics*. Baton Rouge: Louisiana State University Press, 1995.

Clifford, Geraldine. *Those Good Gertrudes: A Social History of Women Teachers in America*. Baltimore: Johns Hopkins University Press, 2014.

Cloward, Richard, and Frances Fox Piven. "A Strategy to End Poverty." *Nation*, May 2, 1966.

Countryman, Matthew. *Up South: Civil Rights and Black Power in Philadelphia*. Philadelphia: University of Pennsylvania Press, 2006.

Cowie, Jefferson. *Capital Moves: RCA's Seventy-Year Quest for Cheap Labor*. Ithaca, NY: Cornell University Press, 1999.

———. "Portrait of the Working Class in a Convex Mirror: Toward a History of the Seventies." *Labor: Studies in Working-Class History of the Americas* 2 (2005): 93–102.

———. *Stayin' Alive: The 1970s and the Last Days of the Working Class*. New York: New Press, 2010.

———. "'Vigorously Left, Right, and Center': The Crosscurrents in Working-Class America in the 1970s." In *America in the Seventies*, edited by Beth Bailey and David Farber, 75–106. Lawrence: University of Kansas Press, 2004.

Critchlow, Donald. *The Conservative Ascendancy: How the GOP Right Made Political History*. Cambridge, MA: Harvard University Press, 2007.

Davis, Mike. *Prisoners of the American Dream: Politics and Economy in the History of the U.S. Working Class.* London: Verso, 1986.

Donovan, John. "A Tale of Two Strikes: The Formation of United Teachers–Los Angeles and the Los Angeles Teacher Strikes of 1970 and 1989." *Southern California Quarterly* 81 (1999): 377–96.

Duménil, Gérard, and Dominique Lévy. *Capital Resurgent: Roots of the Neoliberal Revolution.* Translated by Derek Jeffers. Cambridge, MA: Harvard University Press, 2004.

Ebenstein, Lanny. *Milton Friedman: A Biography.* New York: Palgrave Macmillan, 2007.

Fink, Leon, and Brian Greenberg. *Upheaval in the Quiet Zone: 1199 SEIU and the Politics of Health Care Unionism,* 2nd ed. Urbana: University of Illinois Press, 2009.

Formisano, Ronald. *Boston against Busing: Race, Class, and Ethnicity in the 1960s and 1970s.* Chapel Hill: University of North Carolina Press, 1991.

Fraser, Steve. *The Age of Acquiescence: The Life and Death of American Resistance to Organized Wealth and Power.* New York: Little, Brown, 2015.

Fraser, Steve, and Gary Gerstle, eds. *The Rise and Fall of the New Deal Order, 1930–1980.* Princeton, NJ: Princeton University Press, 1990.

Freeman, Joshua. *In Transit: The Transport Workers Union in New York City, 1933–1966,* 2nd ed. Philadelphia: Temple University Press, 2001.

———. *Working-Class New York: Life and Labor since World War II.* New York: New Press, 2001.

Freund, David. *Colored Property: State Policy and White Racial Politics in Suburban America.* Chicago: University of Chicago Press, 2007.

Gaffney, Dennis. *Teachers United: The Rise of New York State United Teachers.* Albany: State University of New York Press, 2007.

Golin, Steve. *The Newark Teacher Strikes: Hope on the Line.* New Brunswick, NJ: Rutgers University Press, 2002.

Gordon, Colin. *Mapping Decline: St. Louis and the Fate of the American City.* Philadelphia: University of Pennsylvania Press, 2008.

Gordon, Jane Anna. *Why They Couldn't Wait: A Critique of the Black-Jewish Conflict over Community Control in Ocean Hill–Brownsville (1967–1971).* New York: RoutledgeFalmer, 2001.

Gross, James. *Broken Promise: The Subversion of U.S. Labor Relations Policy, 1947–1994.* Philadelphia: Temple University Press, 1995.

Halle, David. *America's Working Man: Work, Home, and Politics among Blue-Collar Property Owners.* Chicago: University of Chicago Press, 1984.

Harvey, David. *A Brief History of Neoliberalism.* Oxford: Oxford University Press, 2005.

———. *The New Imperialism.* New York: Oxford University Press, 2003.

Highsmith, Andrew. *Demolition Means Progress: Flint, Michigan, and the Fate of the American Metropolis.* Chicago: University of Chicago Press, 2015.

Honey, Michael. *Going Down Jericho Road: The Memphis Strike, Martin Luther King's Last Campaign*. New York: W. W. Norton, 2008.

Hower, Joseph. "Jerry Wurf, the Rise of AFSCME, and the Fate of Labor Liberalism, 1947–1981." PhD diss., Georgetown University, 2013.

Jacobs, Meg. *Pocketbook Politics: Economic Citizenship in Twentieth-Century America*. Princeton, NJ: Princeton University Press, 2005.

Johnston, Paul. *Success While Others Fail: Social Movement Unionism and the Public Workplace*. Ithaca, NY: ILR Press, 1994.

Jones, William. *The March on Washington: Jobs, Freedom, and the Forgotten History of Civil Rights*. New York: W. W. Norton, 2013.

Kahlenberg, Richard. *Tough Liberal: Albert Shanker and the Battles over Schools, Unions, Race, and Democracy*. New York: Columbia University Press, 2007.

Kaye, Harvey. *The Fight for the Four Freedoms: What Made FDR and the Greatest Generation Truly Great*. New York: Simon and Schuster, 2014.

Kazin, Michael. *The Populist Persuasion: An American History*. Ithaca, NY: Cornell University Press, 1998.

Kessler-Harris, Alice. *In Pursuit of Equity: Women, Men, and the Quest for Economic Citizenship in Twentieth-Century America*. New York: Oxford University Press, 2001.

Kosner, Edward. *It's News to Me: The Making and Unmaking of an Editor*. New York: Thunder's Mouth Press, 2006.

Kuttner, Robert. *Revolt of the Haves: Tax Rebellion and Hard Times*. New York: Simon and Schuster, 1980.

Lassiter, Matthew. *The Silent Majority: Suburban Politics in the Sunbelt South*. Princeton, NJ: Princeton University Press, 2006.

Levitan, Sar, and Martha Cooper. *Business Lobbies: The Public Good and the Bottom Line*. Baltimore: Johns Hopkins University Press, 1984.

Lewis, Heather. *New York City Public Schools from Brownsville to Bloomberg: Community Control and Its Legacy*. New York: Teachers College Press, 2013.

Lichten, Eric. *Class, Power, and Austerity: The New York City Fiscal Crisis*. South Hadley, MA: Bergin and Garvey, 1986.

Lichtenstein, Nelson. *State of the Union: A Century of American Labor*. Princeton, NJ: Princeton University Press, 2002.

———. *Walter Reuther: The Most Dangerous Man in Detroit*. New York: Basic Books, 1995.

Lieberman, Myron. *The Teacher Unions: How the NEA and the AFT Sabotage Reform and Hold Students, Parents, Teachers, and Taxpayers Hostage to Bureaucracy*. New York: Free Press, 1997.

Lipman, Pauline. *The New Political Economy of Urban Education: Neoliberalism, Race, and the Right to the City*. New York: Routledge, 2011.

Lipset, Seymour Martin, ed. *Unions in Transition: Entering the Second Century*. San Francisco: ICS Press, 1986.

Lortie, Dan. *Schoolteacher: A Sociological Study.* Chicago: University of Chicago Press, 1975.

Lukas, Anthony. *Common Ground: A Turbulent Decade in the Lives of Three American Families.* New York: Vintage, 1985.

Lyons, John. *Teachers and Reform: Chicago and Public Education, 1929–70.* Urbana: University of Illinois Press, 2008.

Mattson, Kevin. *When America Was Great: The Fighting Faith of Postwar Liberalism.* London: Routledge, 2004.

McCartin, Joseph. "Bringing the State's Workers In: Time to Rectify an Imbalanced US Labor Historiography." *Labor History* 47 (2006): 73–94.

———. *Collision Course: Ronald Reagan, the Air Traffic Controllers, and the Strike That Changed America.* New York: Oxford University Press, 2011.

———. "'Fire the Hell Out of Them': Sanitation Workers' Struggles and the Normalization of the Striker Replacement Strategy in the 1970s." *Labor: Studies in Working-Class History of the Americas* 2 (Fall 2005): 67–92.

———. *Labor's Great War: The Struggle for Industrial Democracy and the Origins of Modern Labor Relations, 1912–1921.* Chapel Hill: University of North Carolina Press, 1998.

———. "Turnabout Years: Public Sector Unionism and Fiscal Crisis." In *Rightward Bound: Making America Conservative in the 1970s,* edited by Bruce Schulman and Julian Zelizer, 210–26. Cambridge, MA: Harvard University Press, 2008.

———. "'A Wagner Act for Public Employees': Labor's Deferred Dream and the Rise of Conservatism, 1970–1976." *Journal of American History* 95 (2008): 123–48.

McGirr, Lisa. *Suburban Warriors: The Origins of the New American Right.* Princeton, NJ: Princeton University Press, 2002.

Mertz, Adam. "The 1974 Hortonville Teacher Strike and the Public Sector Labor Dilemma." *Wisconsin Magazine of History* 98 (Spring 2015): 2–13.

Metzgar, Jack. *Striking Steel: Solidarity Remembered.* Philadelphia: Temple University Press, 2000.

Mirel, Jeffrey. *The Rise and Fall of an Urban School System: Detroit, 1907–81,* 2nd ed. Ann Arbor: University of Michigan Press, 1999.

Moe, Terry. *Special Interest: Teachers Unions and America's Public Schools.* Washington, DC: Brookings Institution, 2011.

Moody, Kim. *From Welfare State to Real Estate: Regime Change in New York City, 1974 to the Present.* New York: New Press, 2007.

———. *An Injury to All: The Decline of American Unionism.* London: Verso Press, 1988.

Moreton, Bethany. "Make Payroll, Not War: Business Culture as Youth Culture." In *Rightward Bound: Making America Conservative in the 1970s,* edited by Bruce Schulman and Julian Zelizer, 52–70. Cambridge, MA: Harvard University Press, 2008.

———. *To Serve God and Wal-Mart: The Making of Christian Free Enterprise.* Cambridge, MA: Harvard University Press, 2009.

Mumford, Kevin. *Newark: A History of Race, Rights, and Riots in America*. New York: New York University Press, 2007.

Murphy, Marjorie. *Blackboard Unions: The AFT and the NEA, 1900–1980*. Ithaca, NY: Cornell University Press, 1990.

———. "Militancy in Many Forms: Teachers Strikes and Urban Insurrection, 1967–1974." In *Rebel Rank and File: Labor Militancy and Revolt from Below in the Long 1970s*, edited by Aaron Brenner, Robert Brenner, and Cal Winslow, 229–50. London: Verso, 2010.

Nadasen, Premilla. *Welfare Warriors: The Welfare Rights Movement in the United States*. New York: Routledge, 2005.

Nissenson, Marilyn. *The Lady Upstairs: Dorothy Schiff and the "New York Post."* New York: St. Martin's Press, 2007.

O'Connor, Alice. "Financing the Counterrevolution." In *Rightward Bound: Making America Conservative in the 1970s*, edited by Bruce Schulman and Julian Zelizer, 148–68. Cambridge, MA: Harvard University Press, 2008.

Paige, Rod. *The War against Hope: How Teachers' Unions Hurt Children, Hinder Teachers, and Endanger Public Education*. Nashville, TN: Thomas Nelson, 2006.

Peck, Jamie. *Constructions of Neoliberalism Reason*. Oxford: Oxford University Press, 2010.

Perlstein, Daniel. *Justice, Justice: School Politics and the Eclipse of Liberalism*. New York: Peter Lang, 2004.

Perlstein, Rick. *Before the Storm: Barry Goldwater and the Unmaking of the American Consensus*. New York: Hill and Wang, 2001.

———. *The Invisible Bridge: The Fall of Nixon and the Rise of Reagan*. New York: Simon and Schuster, 2014.

———. *Nixonland: The Rise of a President and the Fracturing of America*. New York: Scribner, 2008.

Perrillo, Jonna. *Uncivil Rights: Teachers, Unions, and Race in the Battle for School Equity*. Chicago: University of Chicago Press, 2012.

Phillips-Fein, Kim. *Invisible Hands: The Making of the Conservative Movement from the New Deal to Reagan*. New York: W. W. Norton, 2009.

Piketty, Thomas. *Capital in the Twenty-First Century*. Cambridge, MA: Harvard University Press, 2014.

Podair, Jerald. *The Strike That Changed New York: Blacks, Whites, and the Ocean Hill–Brownsville Crisis*. New Haven, CT: Yale University Press, 2004.

Quadagno, Jill. *The Color of Welfare: How Racism Undermined the War on Poverty*. New York: Oxford University Press, 1994.

Ravitch, Diane. *Reign of Error: The Hoax of the Privatization Movement and the Danger to America's Public Schools*. New York: Knopf, 2013.

Reese, Ellen. *Backlash against Welfare Mothers: Past and Present*. Berkeley: University of California Press, 2005.

Reese, William. *America's Public Schools: From the Common School to "No Child Left Behind,"* rev. ed. Baltimore: Johns Hopkins University Press, 2011.

Robertson, Susan. "'Remaking the World': Neoliberalism and the Transformation of Education and Teachers' Labor." In *The Global Assault on Teaching, Teachers, and Their Unions: Stories for Resistance,* edited by Mary Compton and Lois Weiner, 11–28. New York: Palgrave Macmillan 2008.

Rousmaniere, Kate. *Citizen Teacher: The Life and Leadership of Margaret Haley.* Albany: SUNY Press, 2005.

Rubio, Philip. *There's Always Work at the Post Office: African American Postal Workers and the Fight for Jobs, Justice, and Equality.* Chapel Hill: University of North Carolina Press, 2010.

Ryan, Francis. *AFSCME's Philadelphia Story: Municipal Workers and Urban Power in the Twentieth Century.* Philadelphia: Temple University Press, 2010.

Schulman, Bruce. *The Seventies: The Great Shift in American Culture, Society, and Politics* Boston: Da Capo Press, 2001.

Self, Robert. *All in the Family: The Realignment of American Democracy since the 1960s.* New York: Hill and Wang, 2012.

———. *American Babylon: Race and the Struggle for Postwar Oakland.* Princeton, NJ: Princeton University Press, 2003.

Sharp, Joanne. *Condensing the Cold War: "Reader's Digest" and American Identity.* Minneapolis: University of Minnesota Press, 2000.

Shelton, Jon. "Letters to the Essex County Penitentiary: David Selden and the Fracturing of America." *Journal of Social History* 48 (Fall 2014): 135–55.

Shermer, Elizabeth Tandy. *Sunbelt Capitalism: Phoenix and the Transformation of American Politics.* Philadelphia: University of Pennsylvania Press, 2013.

Shipps, Dorothy. *School Reform, Corporate Style: Chicago, 1800–2000.* Lawrence: University of Kansas Press, 2006.

Squires, James. *Read All about It! The Corporate Takeover of America's Newspapers.* New York: Times Books, 1993.

Stein, Judith. *Pivotal Decade: How the United States Traded Factories for Finance in the Seventies.* New Haven, CT: Yale University Press, 2010.

Sterling, Christopher, and Timothy Haight. *The Mass Media: Aspen Institute Guide to Communication Industry Trends.* New York: Praeger, 1978.

Stromquist, Shelton. *Reinventing "The People": The Progressive Movement, the Class Problem, and the Origins of Modern Liberalism.* Urbana: University of Illinois Press, 2006.

Sugrue, Thomas. *The Origins of the Urban Crisis: Race and Inequality in Postwar Detroit,* 2nd ed. Princeton, NJ: Princeton University Press, 2005.

Taylor, Clarence. *Reds at the Blackboard: Communism, Civil Rights, and the New York City Teachers Union.* New York: Columbia University Press, 2011.

Tifft, Susan, and Alex Jones. *The Trust: The Private and Powerful Family behind the "New York Times."* Boston: Little, Brown, 1999.

Trotter, Joe, and Jared Day. *Race and Renaissance: African Americans in Pittsburgh since World War II.* Pittsburgh: University of Pittsburgh Press, 2010.

Uetricht, Micah. *Strike for America: Chicago Teachers against Austerity.* London: Verso, 2014.

Vinel, Jean-Christian. *The Employee: A Political History.* Philadelphia: University of Pennsylvania Press, 2013.

White, Richard. *Railroaded: The Transcontinentals and the Making of Modern America.* New York: W. W. Norton, 2011.

Williams, Mary Frase. *The Public School and Finances: A Guide to the Issues in Financing Public Education in the United States.* New York City: Pilgrim Press, 1980.

Witwer, David. "The Heyday of the Labor Beat." *Labor: Studies in Working-Class History of the Americas* 10 (2013): 9–29.

———. *Shadow of the Racketeer: Scandal in Organized Labor.* Urbana: University of Illinois Press, 2009.

Wong, Kenneth. *Funding Public Schools: Politics and Policies.* Lawrence: University Press of Kansas, 1999.

Woodard, Komozi. *A Nation within a Nation: Amiri Baraka (LeRoi Jones) and Black Power Politics.* Chapel Hill: University of North Carolina Press, 1999.

Zaretsky, Natasha. *No Direction Home: The American Family and the Fear of National Decline, 1968–1980.* Chapel Hill: University of North Carolina Press, 2007.

INDEX

JON SHELTON is an assistant professor of democracy and justice studies at the University of Wisconsin Green Bay.

The University of Illinois Press
is a founding member of the
Association of American University Presses.

—————————————————————————

Composed in 10/13 Sabon LT Std
by Lisa Connery
at the University of Illinois Press
Cover designed by Jim Proefrock
Cover illustration: ©iStock.com/Liliboas

University of Illinois Press
1325 South Oak Street
Champaign, IL 61820-6903
www.press.uillinois.edu